Algebra I
Common Core Regents
Course Workbook

Donny Brusca

www.CourseWorkbooks.com

Algebra I
Common Core Regents
Course Workbook

Donny Brusca

First Edition

ISBN 978-1500575267

www.CourseWorkbooks.com

Table of Contents

Introduction

ABOUT THIS BOOK

Every topic section begins with an explanation of the <u>Key Terms and Concepts</u>. Every effort has been made to limit the content here to the most essential ideas. The notes are intended to supplement a fuller presentation of the concepts by the teacher, perhaps through a more developmental approach. The brevity of these notes does not suggest that teachers should "teach to the test," but at a minimum, it is hoped that these notes may sufficiently do so.

Sprinkled liberally throughout the book, denoted by a calculator icon as shown to the left, are instructions on how to use the TI-83 graphing calculator to solve problems or check solutions. Keystrokes include button names in rectangles, STO▸, alternate button features in brackets, [SIN⁻¹], on-screen text in larger rectangles, NUM , and numbers or other mathematical symbols. Directions for selecting on-screen text (arrow keys and ENTER) are usually omitted.

Topic sections include one or more <u>Model Problems</u>, each with a solution and an explanation of steps needed to solve the problem. Steps lettered (A), (B), etc., in the explanations refer to the corresponding lettered steps shown in the solutions. General wording is used in the explanations so that students may apply the steps directly to new but similar problems. However, for clarity, the text often refers to the specific model problem by using *[italicized text in brackets]*. To make the most sense of this writing style, insert the words "in this case" before reading any *[italicized text in brackets]*.

After the Model Problem are a number of <u>Practice Problems</u> in boxed work spaces. These numbered problems are arranged in order of increasing difficulty.

After the Practice Problems are <u>Regents Questions</u> on the topic, all of which have appeared on past New York State Regents exams, including *every* Algebra I Common Core exam question (as of this book's publication date), plus a number of questions from previous Regents exam formats, including Integrated Algebra, Algebra II / Trigonometry, Math A and B, Sequential Math, Ninth to Eleventh Year Math, and even the earlier Algebra exams. The sampler questions from the May 2013 Common Core publication are also included. Questions are grouped by *Multiple Choice* questions followed by *Constructed Response* questions, generally in chronological order of appearance.

At the end of the book are a *Sample Regents Exam* and a *Concordance to Common Core Standards*. An Answer Key is available at <u>www.CourseWorkbooks.com</u> and is free to organizations that purchase class sets.

ABOUT THE AUTHOR

Donny Brusca has taught for about 25 years, mostly on the high school and college levels. He is currently employed as the Director of Student Data Management at the Williamsburg Charter High School in Brooklyn. He has a B.S. in mathematics and M.A. in computer and information science from Brooklyn College (CUNY) and a post-graduate P.D. in educational administration and supervision from St. John's University. For more information about the author, visit <u>www.Brusca.info</u>.

I.   PRE-ALGEBRA REVIEW

Number Sets

Counting numbers (also called **Natural numbers**): $\{1, 2, 3, 4, 5, \ldots\}$
Whole numbers include the counting numbers and zero: $\{0, 1, 2, 3, 4, 5, \ldots\}$
Integers include the whole numbers and their opposites: $\{\ldots -3, -2, -1, 0, 1, 2, 3, \ldots\}$

Rational numbers can be expressed as $\frac{a}{b}$ where a and b are integers and $b \neq 0$.
Every rational number can be expressed as a **terminating or repeating decimal**.

Irrational numbers are all the real numbers that are *not* rational; that is, they cannot be expressed as a quotient of integers, and their decimals are non-repeating and non-terminating. π is an irrational number. 3.14 is an *approximation* of π, rounded to the nearest hundredth.

The **Real numbers** include all the rational and irrational numbers, and are represented by all the points that make up a number line or a coordinate axis.

The subsets of the Real numbers are demonstrated below:

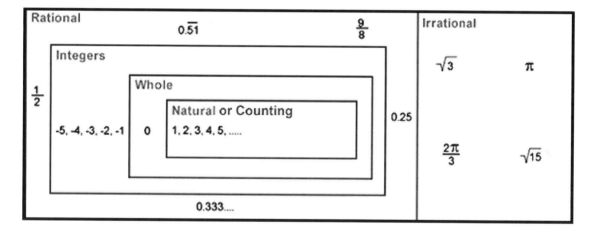

Set Notation

A **set** is a collection of objects.

Roster form lists the elements of a set inside braces { }.
For examples: {2, 3, 4} or {John, Paul, George, Ringo}

In roster form, **ellipses** (...) are often used to show that a pattern continues.
For examples: The set of counting numbers can be written as {1, 2, 3,...}
                      The set of integers can be written as {..., –3, –2, –1, 0, 1, 2, 3,...}

A **finite set** has a certain number of elements; an **infinite set** does not.
For examples: The set of counting numbers less than 6, {1, 2, 3, 4, 5}, is finite because it has 5
                      elements, but the set of counting numbers, {1, 2, 3,...}, is infinite

An **empty set** (aka Null set) is a set with no elements, symbolized by { } or $\emptyset$;
For example: The set of even prime numbers greater than 2 = { }

Interval notation uses parentheses ( ) and/or brackets [ ] to name the endpoints (lower and upper
bounds) of a set of all real numbers between those endpoints.
   • A parenthesis represents an "open" endpoint ($>$ or $<$; not included in the set)
   • A bracket represents a "closed" endpoint ($\geq$ or $\leq$; included in the set)
These correspond to the open and closed circles in the graph of an inequality on a number line.
For example, $(-1, 5]$ represents all real numbers x such that $-1 < x \leq 5$.

If a set has no upper bound, the **infinity symbol** ∞ is used; for no lower bound, $-\infty$ is used. In
either case, since there is no endpoint, use a parenthesis.
For examples: $[3, \infty)$ for $x \geq 3$ $(-\infty, 2)$ for $x < 2$ $(-\infty, \infty)$ for all real numbers

Inclusive means that the endpoints are included (closed); **exclusive** means they are not (open).
For example: The set of integers between 3 and 10, inclusive, is [3, 10].

Set-builder notation: uses braces { } and a vertical bar | to define a set by the properties that its
members must satisfy. It will often start with "{x|}" which is read as "the set of all x such that."
For example: {3, 4, 5} can be written as {x| $3 \leq x \leq 5$, where x is a whole number}

An alternate format of set-builder notation places the number type immediately after the variable.
For example: {3, 4, 5} can also be written as {x whole | $3 \leq x \leq 5$ }

The **inequality symbols** used are important: $\leq$ is closed (*inclusive*); $<$ is open (*exclusive*).
For example: {3, 4, 5} can also be written as { x whole | $2 < x < 6$ }

Order of Operations

Order of Operations:
1. Exponents
2. Multiplication and Division (left to right)
3. Addition and Subtraction (left to right)

Note that steps 2 and 3 are performed **left to right**.

For examples: $6 \div 3 \times 2$ results in 4 (perform the division first, going left to right)

$6 - 3 + 2$ results in 5 (perform the subtraction first, going left to right)

Parentheses (or brackets, absolute value symbols, fraction bars, radical signs, etc.) can change the normal order of operations.

For examples: $3 \times (4 + 5)$ forces the addition to be performed before the multiplication

$(2x)^2$ means square the product, not just the x

When an expression has **multiple layers** of parentheses or brackets, work from within the innermost first.

For example: $2 \times [9 + (5 - 2)] = 2 \times [9 + 3] = 2 \times 12 = 24$

 When entering expressions with a **fraction bar** into a calculator, be sure to enter the numerator in parentheses and the denominator in parentheses.

For example: $\dfrac{9+3}{3+1}$ should be entered as $(9 + 3) \div (3+1)$, resulting in $12/4 = 3$.

If the parentheses are omitted, $9 + 3 \div 3 + 1$ would result in $9 + 1 + 1 = 11$.

Integers

Absolute Value: the distance that a number is from zero, written using the vertical symbols, | |
For examples: |5| is 5 |-8| is 8 |0| is 0

On the calculator, use MATH NUM abs(for absolute value.

Multiplying and Dividing Integers:
    Same Signs → Positive result
    Different Signs → Negative result

Examples: $(-5)(-2) = 10$ (same signs) $(5)(-2) = -10$ (different signs)

$$\frac{-12}{-3} = 4 \qquad \text{(same signs)} \qquad\qquad \frac{-12}{3} = -4 \qquad \text{(different signs)}$$

Adding and Subtracting Integers:
    Same Signs → Add the absolute values, and keep the same sign
    Different Signs → Subtract the absolute values, and keep the sign of the larger

"Terms" Method for Addition and Subtraction of Integers
Eliminate "Double Signs" and rewrite without parentheses around integers, then add terms.
Double Sign Rules: $a + (+b) = a + b$
                       $a + (-b) = a - b$
                       $a - (+b) = a - b$
                       $a - (-b) = a + b$
Examples:

$(+3) + (+2)$	$= 3 + 2 = 5$	$(-3) + (-2)$	$= -3 - 2 = -5$
$(+3) + (-2)$	$= 3 - 2 = 1$	$(-3) + (+2)$	$= -3 + 2 = -1$
$(+3) - (+2)$	$= 3 - 2 = 1$	$(-3) - (+2)$	$= -3 - 2 = -5$
$(+3) - (-2)$	$= 3 + 2 = 5$	$(-3) - (-2)$	$= -3 + 2 = -1$

Be careful about the **Order of Operations**:
If there is an *operation in parentheses*, it must be calculated before adding or subtracting terms.
For example: $10 - (-8 + 6)$ $= 10 - (-2)$ $= 10 + 2 = 12$
Exponents, multiplication and division are also done before adding or subtracting terms.
For example: $5 - (-2)^2$ $= 5 - 4 = 1$

Adding and subtracting a series of terms:
                                                    (A)   (B)   (C)
        $2 + (-3) - (-4) + 5 + (-6)$ $= 2 \underline{-3} + 4 + 5 \underline{-6}$ $= 11 \underline{-9} = 2$

Explanation of steps:
    (A) Use the "Double Sign" rules to eliminate the parentheses.
    (B) Add the positive terms *[to get 11]* and the negative (underlined) terms *[to get –9]*.
    (C) Use the Addition of Integers rules to get the result.

Prime Factorization

Prime numbers: whole numbers <u>greater than 1</u> that have exactly two factors, 1 and itself.
>   The first 5 prime numbers are: 2, 3, 5, 7, 11

Composite numbers: whole numbers <u>greater than 1</u> that are not prime.

Divisibility Rules: a number is divisible by
2	if the last digit is even (divisible by 2)
3	if the *sum of the digits* is divisible by 3
5	if the last digit is 0 or 5

Prime factorization: composite numbers can be expressed uniquely as a product of its prime factors. Use repeated division on the calculator, starting with 2. The end branches are the prime factors. For example, the prime factorization tree for 1,050 is:

1,050	1,050 is even, so divide by 2
/ \	
2 × 525	525 is odd, but divisible by 3 (5+2+5=12)
/ \	
3 × 175	175 is not divisible by 3, but by 5
/ \	
5 × 35	35 is also divisible by 5
∧	
5 × 7	prime factorization is $2 \cdot 3 \cdot 5 \cdot 5 \cdot 7$
	or $2 \cdot 3 \cdot 5^2 \cdot 7$

Using prime factorization to reduce fractions

For example, to reduce $\dfrac{315}{1,050}$:

$$\underset{(A)}{\frac{315}{1,050} = \frac{3 \cdot 3 \cdot 5 \cdot 7}{2 \cdot 3 \cdot 5 \cdot 5 \cdot 7}} \qquad \underset{(B)}{\frac{\cancel{3} \cdot 3 \cdot \cancel{5} \cdot \cancel{7}}{2 \cdot \cancel{3} \cdot \cancel{5} \cdot 5 \cdot \cancel{7}}} = \underset{(C)}{\frac{3}{2 \cdot 5}} = \underset{(D)}{\frac{3}{10}}$$

Explanation of steps:
>   (A) Write the prime factorization of the numerator and denominator.
>   (B) Cancel pairs of common factors appearing in both the numerator and denominator.
>   (C) Write the remaining factors (or if all factors are canceled, write 1).
>   (D) Multiply.

 You can check the result on a calculator: 315 ÷ 1050 [MATH] [Frac]

GCF and LCM

The **GCF (greatest common factor)** is the largest common factor of a set of numbers.
(The GCF is also known as GCD or greatest common divisor.)

Use prime factorization to find the GCF.
From the prime factorizations, the GCF is the product of the common prime factors.
For example: To find the GCF of 27 and 36, we can write the prime factorizations.

$$27 = 3^3 \qquad 36 = 2^2 \cdot 3^2$$

Then write out the *common* bases *[only 3]*, to their *lowest* powers *[2]*.

$$3^2 = 9 \text{ is the GCF.}$$

This method will help us later when factoring out the GCF of an algebraic expression.

You can also find the GCF using the calculator: $\boxed{\text{MATH}}\ \boxed{\text{NUM}}\ \boxed{\text{gcd(}}\ 27\ ,\ 36\ )$

We can use the GCF to reduce fractions.
For example: $\dfrac{27}{36} = \dfrac{27 \div 9}{36 \div 9} = \dfrac{3}{4}$ Divide the numerator and denominator by the GCF.

The **LCM (least common multiple)** is the smallest common multiple of a set of numbers.
(The LCM is also known as LCD or least common denominator when we are referring to the denominators of fractions.)

Use prime factorization to find the LCM.
From the prime factorizations of the numbers, the product of all the prime factors to their highest powers will result in the LCM.
For example: To find the LCM of 27 and 36, we can write the prime factorizations.

$$27 = 3^3 \qquad 36 = 2^2 \cdot 3^2$$

Then write out *all* the prime bases *[2 and 3]*, each to their *highest* powers.

$$2^2 \cdot 3^3 = 108 \text{ is the LCM.}$$

We use the LCM to add fractions. We also use it to solve equations involving fractions.

You can also find the LCM using the calculator: $\boxed{\text{MATH}}\ \boxed{\text{NUM}}\ \boxed{\text{lcm(}}\ 27\ ,\ 36\ )$

Place Values

Every digit in a number has a **place value**.

For example: The following chart shows the place values for the digits of the number 1,234.

$$\begin{array}{cccc} \text{thousands } (10^3) & \text{hundreds } (10^2) & \text{tens } (10^1) & \text{ones } (10^0) \\ 1, & 2 & 3 & 4 \end{array}$$

This means that $1,234 = (1 \times 10^3) + (2 \times 10^2) + (3 \times 10^1) + (4 \times 10^0)$

(Note: $10^0 = 1$)

This also holds true for the digits to the **right of the decimal point**.

For example: The place values for the digits of 0.56789 are shown below.

$$\begin{array}{cccccc} & \text{tenths } (10^{-1}) & \text{hundredths } (10^{-2}) & \text{thousandths } (10^{-3}) & \text{ten-thousandths } (10^{-4}) & \text{hundred-thousandths } (10^{-5}) \\ 0. & 5 & 6 & 7 & 8 & 9 \end{array}$$

So, $0.56789 = (5 \times 10^{-1}) + (6 \times 10^{-2}) + (7 \times 10^{-3}) + (8 \times 10^{-4}) + (9 \times 10^{-5})$

$$= (5 \times \frac{1}{10}) + (6 \times \frac{1}{100}) + (7 \times \frac{1}{1,000}) + (8 \times \frac{1}{10,000}) + (9 \times \frac{1}{100,000})$$

If we place an imaginary mirror over the ones place, the names of the place values to the right of the mirror are the same as their reflections to the left except that they end in "ths".

Note that the first place to the right of the decimal point is the "tenths" place.

Also note that the powers of 10 continue to decrease into the negative powers as we move to the right of the decimal point. 10^{-n} is the reciprocal of 10^n.

Rounding

Rounding to a place left of the decimal point.

Model Problem 1:   Round $5,236,174$ to the *nearest hundred thousand.*

Solution: $5,200,000$
  (A) Determine which digit is in the specified place.
      [The 2 is in the hundred thousands place.]
  (B) Look one place to the right. If the next digit is at least 5, increase the digit at the
      specified place by one; if not, keep the digit the same.
      [To the right of the 2 is a 3, so the 2 will remain unchanged.]
  (C) Change each place to the right of the specified place to zeroes.

Model Problem 2:   Round $5,236,174$ to the *nearest hundred.*

Solution: $5,236,200$
  (A) Determine which digit is in the specified place. *[The 1 is in the hundreds place.]*
  (B) Look one place to the right. If the next digit is at least 5, increase the digit at the
      specified place by one; if not, keep the digit the same.
      [To the right of the 1 is a 7, so the 1 will increase by one to a 2.]
  (C) Change each place to the right of the specified place to zeroes.

Rounding to a place right of the decimal point.
Follow the same steps as above except step (C) is changed.

Model Problem 3:   Round 0.5374 to the *nearest hundredth.*

Solution: 0.54
  (A) Determine which digit is in the specified place. *[The 3 is in the hundredths place.]*
  (B) Look one place to the right. If the next digit is at least 5, increase the digit at the
      specified place by one; if not, keep the digit the same.
      [To the right of the 3 is a 7, so the 3 will increase by one to a 4.]
  (C) Do NOT write zeroes or any other digits after the specified place.

Model Problem 4:   Round 6.975 to the *nearest tenth.*

Solution: 7.0
  (A) Determine which digit is in the specified place. *[The 9 is in the tenths place.]*
  (B) Look one place to the right. If the next digit is at least 5, increase the digit at the
      specified place by one; if not, keep the digit the same. *[To the right of the 9 is a 7, so the
      9 will increase by one. However, since 9+1=10 and 10 has two digits, we need to
      change the 9 to a 0 and then carry the 1 to the left, changing the 6 into a 7.]*
  (C) Do NOT write zeroes or any other digits after the specified place.

Precision of Measurements

Real life math problems tend to involve measurements. The **accuracy** of a measurement is its nearness to the true value. The **precision** of a measurement is the degree to which its accuracy is expressed.

It is reasonable to **report measurements** only to a level of *precision* that the tool allows. A ruler marked in millimeters is more precise than a ruler that is marked only in centimeters.

For example: If someone were to weigh an item on a scale that only had markings at every tenth of a pound, it would be reasonable to give a measure of 25.7 or 25.8 pounds. We could not report a measure of 25.726 pounds.

Also, **calculations based on measurements** should only be reported to a precision based on the number of *significant digits* of the least precise measurement that is given.

To determine the **number of significant digits**, count (a) all nonzero digits, (b) all zeros to the right of the decimal point after the last nonzero digit, and (c) all zeros between significant digits. Zeros at the end of whole numbers are not considered significant digits.

For examples: The number of significant digits in
31.25 is 4; 2.20 is 3; 0.00150 is 3; 3050.0 is 5; and 250,000 is 2.

When **adding or subtracting**, round to the *same place as the last significant digit* of the least precise measurement.

For example: 16 cm + 23.6 cm = 39.6 cm should be rounded to the nearest whole number, 40 cm, since 16 cm is the less precise measurement shown to only the units place.

When **multiplying or dividing**, the result should be rounded to the *same number of significant digits* as the measurement with the fewest significant digits.

For example: 16 cm × 23.6 cm = 377.6 cm^2 should be rounded to 380 cm^2, only two significant digits, since 16 cm has only two significant digits.

If it's possible to express an answer as an **exact fraction** instead of a rounded decimal, it is more accurate to do so.

For example: When dividing 17÷3, the calculator shows 5.6666667. Rather than rounding this decimal, it is more accurate to write an answer of $5\frac{2}{3}$.

Model Problem 1: *reporting measurements*
Victor claims to have measured the thickness of a penny with a ruler. Which of the following is most likely the value Victor measured?

    (1) 0.061 in   (3) 1.5 in
    (2) 1.55 mm   (4) 1.5 mm

Solution:
(4) 1.5mm

Explanation:
Use your knowledge of the given units and of the level of precision that the measuring tool allows. *[While (1) and (2) are both the correct measurements for the thickness of a penny, a simple ruler wouldn't be able to deliver that kind of precision. Choice (3) is clearly an incorrect measurement for the thickness of a penny, so the correct answer is (4).]*

<u>Model Problem 2</u>: *significant digits*
How many significant digits are in each of the following?
    (a) 204,000   (b) 2.040   (c) 0.00204

Solution:
    (a) 3   (b) 4   (c) 3

Explanation:
(a) Zeroes at the ends of whole numbers *[the last three zeroes here]* are not significant.
(b) Count zeroes to the right of the decimal point after the last nonzero digit. *[So, both count.]*
(c) Don't count lead zeroes in a decimal. *[Only the last three digits are significant.]*

<u>Model Problem 3</u>: *calculations based on adding measurements*
What is the sum of $6.6412 + 12.85 + 0.046 + 3.48$ grams expressed to the correct number of significant digits?

Solution:
23.02 g.

Explanation:
For a sum, round to the *same place as the last significant digit* of the least precise measurement.
[The least precise measures, 12.85 and 3.48, are given to the hundredths place only, so the sum should be rounded to the same place value.]

<u>Model Problem 4</u>: *calculations based on multiplying measurements*
The dimensions of a rectangle are given as 8.7 by 3.16 inches. The area is calculated by multiplying, giving a result on the calculator of 27.492. To what place should this be rounded?

Solution:
To the units place, 27.

Explanation:
For a product, round to the *same number of significant digits* as the measurement with the fewest significant digits.
[8.7 has only two significant digits, so the result should as well.]

Exponents

A **positive exponent** represents the number of times the base is used as a factor.

For example: $x^5 = x \cdot x \cdot x \cdot x \cdot x$ $x^1 = x$

A base with a **zero exponent** evaluates to 1: $x^0 = 1$

A base with a **negative exponent** evaluates to the reciprocal (*multiplicative inverse*) of the same base to the opposite (*positive*) exponent.

For example: x^{-3} is the reciprocal of x^3, which is $\dfrac{1}{x^3}$.

Addition and Subtraction Rules:

Must be like terms (same base *and* exponent): <u>Keep</u> the same exponent and add or subtract the coefficients

For examples: $x^5 + x^3$ cannot be combined; not like terms

$$x^3 + x^3 = 2x^3$$
$$4x^3 - x^3 = 3x^3$$

Multiplication Rule (with same base): <u>Add</u> the exponents

For example: $x^5 \cdot x^3 = x^{5+3} = x^8$ Why it works: $(x \cdot x \cdot x \cdot x \cdot x)(x \cdot x \cdot x) = x^8$

Division Rule (with same base): <u>Subtract</u> the exponents

For examples: $\dfrac{x^5}{x^3} = x^{5-3} = x^2$ Why it works: $\dfrac{\not x \cdot \not x \cdot \not x \cdot x \cdot x}{\not x \cdot \not x \cdot \not x} = x \cdot x = x^2$

$\dfrac{x^3}{x^3} = x^{3-3} = x^0 = 1$ Why it works: $\dfrac{\not x \cdot \not x \cdot \not x}{\not x \cdot \not x \cdot \not x} = \dfrac{1}{1} = 1$

$\dfrac{x^3}{x^5} = x^{3-5} = x^{-2} = \dfrac{1}{x^2}$ Why it works: $\dfrac{\not x \cdot \not x \cdot \not x}{\not x \cdot \not x \cdot \not x \cdot x \cdot x} = \dfrac{1}{x \cdot x} = \dfrac{1}{x^2}$

Power Rule: <u>Multiply</u> the exponent by the power

For example: $(x^5)^3 = x^{5 \cdot 3} = x^{15}$ Why it works: $(x^5)(x^5)(x^5) = x^{15}$

Rules Summary:

Operation	Exponents
Addition/Subtraction	Keep
Multiplication	Add
Division	Subtract
Raise to a Power	Multiply

A **term** is a number, a variable, or any product or quotient of numbers and variables.
A **monomial** is a single term without any variables in the denominator, such as $-2xy$.

Operations on Monomials:
  a) Perform the given operation on the coefficients
  b) For common bases, apply the proper rule for exponents
For examples:

$2x^2y + 3x^2y = 5x^2y$ <u>Add</u> coefficients $[2+3]$; <u>keep</u> variable parts $[x^2y]$

$(3a^2)(4ac) = 12a^3c$ <u>Multiply</u> coefficients $[3 \times 4]$; <u>add</u> exponents $[a^{2+1}c]$

$\dfrac{30x^6y^3z^2}{2x^4y^3z} = 15x^2z$ <u>Divide</u> coefficients $[30 \div 2]$; <u>subtract</u> exponents $[x^{6-4}y^{3-3}z^{2-1}]$

$(-3a^2b)^3 = -27a^6b^3$ <u>Raise</u> coefficient to the power $[(-3)^3]$; <u>multiply</u> exponents $[a^{2 \cdot 3}b^{1 \cdot 3}]$

When Dividing Monomials: If the variable's **exponent is larger in the denominator** (i.e., the difference is negative), leave the variable in the denominator with a positive exponent.

For example: $\dfrac{8x^2y}{4xy^5} = \dfrac{2x}{y^4}$ y^4 remains in the denominator since $\dfrac{y}{y^5} = y^{1-5} = y^{-4} = \dfrac{1}{y^4}$

<u>Model Problem</u>
What is the product of $3w^2x$ and $(2w^2x^3y)^3$?

Solution:
$$(3w^2x)(2w^2x^3y)^3 =$$
(A) $(3w^2x)(8w^6x^9y^3) =$
(B) $24w^8x^{10}y^3$

Explanation of Steps:
Perform the operations using the normal order of operations, following the rules for exponents.
  (A) *[Here, the second expression involves raising to a power, so this is performed first. Raise $2^3 = 8$ and then use the powers rule to multiplying the exponents by the power $w^{2 \cdot 3}x^{3 \cdot 3}y^3 = w^6x^9y^3$]*
  (B) *[Then multiply the monomials by multiplying the coefficients $3 \times 8 = 24$ and using the multiplication rule to add the exponents $w^{2+6}x^{1+9}y^3 = w^8x^{10}y^3$]*

Scientific Notation

Scientific notation is generally used to write numbers with very large or very small absolute values. The notation uses the product of a decimal and a power of 10. The decimal must have a **single non-zero digit before the decimal point**.

To change from scientific to standard notation, the power of 10 tells you how many places to move the decimal point to the right (if positive) or to the left (if negative).
For examples:

9.3×10^7 Move decimal point 7 places right   93,000,000

2.9×10^{-6} Move decimal point 6 places left   0.0000029

 To enter scientific notation into the calculator, use the [2nd] [EE] keys in place of "× 10".

9.3×10^7 is entered as: 9.3 [2nd] [EE] 7

To change a number into scientific notation:
1. Move the decimal point so that it comes **after the first non-zero digit** in the number.
2. Count how many places we would need to move the decimal point to get back to the original number. If the decimal point needs to move right, the power of 10 is positive; otherwise, the power of 10 is negative.
3. Write as the product of the decimal and the power of 10.

Changing large numbers to scientific notation:

93,000,000   9 3,000,000₀ 9.3×10^7

We'd need to move decimal point 7 places right to get back to the original number.

Changing small numbers to scientific notation:

0.0000029   0₀000002 9 2.9×10^{-6}

We'd need to move decimal point 6 places left to get back to the original number.

 Alternatively, we can set the calculator to display in scientific notation and then enter the value. To have the calculator display all values in scientific notation:
1. Press [MODE] then the right arrow [▶] to select ⌐Sci⌐ and [ENTER]. Press [2nd] [QUIT].
2. Enter the number normally (for example, 0.0000029) and press [ENTER]. The calculator displays 2.9 ᴇ -6, representing 2.9×10^{-6}.
3. *Important:* To return to standard display, press [MODE] ⌐Normal⌐ [ENTER] [2nd][QUIT].

To **multiply or divide** numbers in scientific notation:
1. Multiply or divide the decimals
2. Use the rules for exponents to multiply or divide the powers of 10
3. Adjust the decimal point if necessary, making sure to increase or decrease the power of 10 to compensate for the adjustment (when moving left, increase the power; when moving right, decrease the power)

For example: $\dfrac{(64 \times 10^9)}{(3.2 \times 10^6)} = \dfrac{64}{3.2} \times \dfrac{10^9}{10^6} = 20 \times 10^3 = 2.0 \times 10^4$

We can also use the calculator to perform the operation:
1. Set to "Sci" display: [MODE] [Sci] [ENTER] [2nd] [QUIT]
2. Enter the expression: 64 [2nd] [EE] 9 ÷ 3.2 [2nd] [EE] 6 [ENTER]

Model Problem 1: *converting between normal and scientific notations*
Write 0.000257 in scientific notation. (*For this problem, do not use the calculator.*)

Solution:

(A)   (B)

2.57×10^{-4}

Explanation of Steps:
(A) Move the decimal point to the right of the first non-zero digit *[after the 2]*.
(B) Count how many places we need to move the decimal point to return to the original number, and make that the power of 10. *[We need to move 4 places left, so 10^{-4}.]*

Model Problem 2: *operations*
The mass of a single oxygen molecule (O_2) is approximately 5.356 X 10^{-26} kg. What is the approximate total mass of 5 X 10^{20} oxygen molecules, written in scientific notation?

Solution:

(A) $(5.356 \times 10^{-26})(5 \times 10^{20}) =$

(B) $(5.356 \times 5)(10^{-26} \times 10^{20}) =$

$26.78 \times 10^{-6} =$

(C) 2.678×10^{-5}

Alternate Solution:
On the calculator,

[MODE] [Sci] [ENTER] [2nd] [QUIT]

5.356 [2nd] [EE] [(-)] 26 × 5 [2nd] [EE] 20 [ENTER]

Explanation of Steps:
(A) Write the product or quotient *[since we are given the mass of one molecule and need to calculate the total mass of a large number of molecules, we need to find the product]*.
(B) Calculate the decimal part *[5.356 X 5 = 26.78]* and use the rules for exponents to calculate the power of 10 *[since we are multiplying powers of the same base, we add exponents: –26 + 20 = –6]*.
(C) Adjust the decimal point so that it appears after the first non-zero digit, and compensate the power of 10 accordingly *[moving the decimal point one place left means we have to increase the power of 10 by one, from –6 to –5, to compensate]*.

Fractions

Converting between fractions and decimals

To change a fraction to a decimal:

Divide the numerator by the denominator on a calculator.

For example: $\dfrac{1}{25}$ is entered as $1 \div 25$ [ENTER] resulting in 0.04.

To change a decimal to a fraction:

On the calculator, using [MATH] [Frac]. If the entered value is irrational – for example, $\sqrt{2}$ – there is no equivalent fraction, so it is left as a decimal.

For example: Entering 0.04 [MATH] [Frac] results in 1/25.

Comparing fractions

You can compare two fractions by cross-multiplying and placing each cross product above the fraction whose numerator is used in the product. The fraction with the larger cross product above its numerator is the larger fraction. If the cross products are equal, the fractions are equivalent.

Which is larger, $\dfrac{5}{9}$ or $\dfrac{6}{10}$?

Solution:

(A)   50   54

(B) $\dfrac{5}{9} < \dfrac{6}{10}$

Explanation of steps:

(A) Place the cross product, $5 \times 10 = 50$, above the first fraction, and the cross product, $6 \times 9 = 54$, above the second.

(B) Since $50 < 54$, the first fraction is less than the second fraction.

Multiplying fractions

$$\frac{5}{9} \times \frac{6}{10} \overset{(A)}{=} \frac{\cancel{5}}{\cancel{3} \cdot 3} \times \frac{2 \cdot \cancel{3}}{2 \cdot \cancel{5}} \overset{(B)}{=} \frac{1}{3}$$

Check on calculator: $(5 \div 9) \times (6 \div 10)$ [MATH] [Frac]

Explanation of steps:

(A) Write the prime factorization of each number. Cancel pairs of prime factors which are common to a numerator and denominator of *either* fraction.

(B) Multiply remaining factors across.

<u>Dividing fractions</u>

$$\underset{}{\frac{3}{10} \div \frac{1}{2}} = \overset{(A)}{\frac{3}{10} \times \frac{2}{1}} = \overset{(B)}{\frac{3}{2 \cdot 5} \times \frac{\not{2}}{1}} = \frac{3}{5}$$

Check on calculator: $(3 \div 10) \div (1 \div 2)$ [MATH] [Frac]

Explanation of steps:
    (A) "Flip" the second fraction (change it to its reciprocal)
    (B) Multiply, following the steps above for multiplying fractions

<u>Adding or subtracting fractions</u>

$$\overset{(A)}{\frac{3}{10} + \frac{7}{15} = \frac{3}{2 \cdot 5} + \frac{7}{3 \cdot 5}} \quad \overset{(B)}{LCM = 2 \cdot 3 \cdot 5 = 30} \quad \overset{(C)}{\frac{3 \cdot (3)}{30} + \frac{7 \cdot (2)}{30}} = \overset{(D)}{\frac{9}{30} + \frac{14}{30} = \frac{23}{30}}$$

Check on calculator: $(3 \div 10) + (7 \div 15)$ [MATH] [Frac]

Explanations of steps:
    (A) Change denominators to their prime factorizations.
    (B) Find the LCM of the denominators. *[Remember to use **all** prime factors, each to their **highest** powers; see the previous section on GCF and LCM for details.]* This will be the denominator of the answer.
    (C) Multiply each numerator by whatever prime factors are "missing" in its denominator.
    (D) Add (or subtract) numerators across; the denominator stays the same.

Important Note: In all of the above examples, we could have used the calculator to find the results more quickly. However, when we learn to perform operations on algebraic fractions in Algebra II, the prime factorization methods we use here will work even with variables, so it is beneficial to learn the methods now.

Evaluating Expressions and Formulas

An algebraic **expression** may contain numbers, variables, operations, and other mathematical symbols such as parentheses.

For example: $\qquad x-(y+1)$

To evaluate an algebraic expression:
1. Rewrite the expression by replacing each variable with its value in parentheses.
2. Evaluate using the correct order of operations.

The **absolute value** of a number n is the distance between n and 0, written using the vertical symbols, $|n|$. The absolute value of a positive number (or 0) is the number itself; the absolute value of a negative number is its opposite.

For examples: $\quad |5|$ is 5 $\qquad |-8|$ is 8 $\quad |0|$ is 0

On the calculator, use [MATH] [NUM] [abs(] for absolute value.

Model Problem

What is the value of $3x - y^2$ when $x = 5$ and $y = -3$?

Solution:	**Explanation of Steps:**
$3x - y^2 =$	(A) Rewrite the expression by replacing each variable with its value in parentheses.
(A) $\quad 3(5) - (-3)^2 =$	*[Replace x with (5) and y with (–3).]*
(B) $\qquad 3(5) - 9 =$	(B) Evaluate using the correct order of operations *[the exponent, then the multiplication, then the subtraction]*.
$\qquad 15 - 9 = 6$	

Check using the calculator:

1. Store the values of the variables: Display

        5 [STO▸][ALPHA] [X] [ENTER]   5 → X

        [(-)]3 [STO▸][ALPHA] [Y] [ENTER]   –3 → Y

2. Enter the expression:

        3 [ALPHA] [X] – [ALPHA] [Y] [x²] [ENTER]6

A **formula** is an equation with a single variable on one side and an expression involving another variable (or variables) on the other.

For examples: $\quad F = \dfrac{9}{5}C + 32$ finds the Fahrenheit (F) temperature for a given Celsius (C).

$\qquad\qquad A = \dfrac{bh}{2}$ finds the Area of a triangle (A) given its base (b) and height (h).

To evaluate a formula:
1. Substitute the given value(s) for the appropriate variable(s).
2. Evaluate the expression to find the value of the desired variable.

For example: To convert $-10°$ Celsius (C) to degrees Fahrenheit (F),

$$F = \frac{9}{5}(-10) + 32$$

$$F = -18 + 32 = 14 \qquad \text{So, } -10°C = 14°F$$

Solving Simple Equations

An **equation** is a statement that one expression is equal to another. It contains an = sign.
For example: $3x - 1 = x + 5$

A **variable term** in an equation includes the variable as a factor.
A **constant term** in an equation does not include a variable factor.
For example: $3x + 5 = 35$ has a variable term $[3x]$ and a constant term $[+5]$ on the left side.

The goal when solving an equation is to **isolate the variable** (transform it into $x = a$ *value*). Do this by using the <u>reverse order of operations</u>:
   (a) add the opposite (additive inverse) of the constant term to both sides.
   (b) divide both sides by (or multiply both sides by the reciprocal, or multiplicative inverse, of) the variable term's coefficient.

To check your solution:
Substitute your solution for the variable in the original equation. *It is usually best to use parentheses around the value when substituting.* Then, evaluate both sides of the equation to determine whether the solution makes the equation true.

<u>Model Problem 1</u>: *one-step equations*
Solve for x: $x - 6 = 12$

Solution:

$$x - 6 = 12$$
$$\underline{+6 \quad +6}$$
$$x \quad = 18$$

Check:

$$x - 6 = 12$$
$$(18) - 6 = 12$$
$$12 = 12 \checkmark$$

Explanation:
To isolate the variable *[x]*, we need to eliminate anything else from the same side of the equation *[–6]*. We do this by performing the inverse operation *[+6]* to both sides of the equation.

<u>Model Problem 2</u>: *two-step equations*
Solve for x: $3x + 5 = 35$

Solution:

$$3x + 5 = 35$$
$$\text{(A)} \quad \underline{-5 \quad -5}$$
$$\text{(B)} \quad \frac{3x}{3} = \frac{30}{3}$$
$$x = 10$$

Check:

$$3(10) + 5 = 35$$
$$35 = 35 \checkmark$$

Explanation of steps:
   (A) Eliminate the constant term *[eliminate +5 by adding –5 to both sides]*.
   (B) Eliminate the coefficient of the variable term *[eliminate the 3 by dividing both sides by 3]*.

Rates

A **ratio** is a comparison, by division, of two quantities.

The ratio of "*a* to *b*" can be expressed as $\dfrac{a}{b}$ or $a:b$.

A ratio can be **simplified** by dividing each quantity by the GCF, if the GCF > 1.

For example: The sides of a rectangle measure 40 inches and 15 inches. The sides are in the ratio 40:15. Since the GCF of 40 and 15 is 5, this ratio can be simplified to 8:3.

A **rate** is a special type of ratio involving two quantities measured in different units.

For example: A rate of 60 miles in 3 hours $= \dfrac{60\,miles}{3\,hours}$

A **unit rate** is a rate that contains a unit measure (1) in the denominator. A unit rate is usually expressed with the joining word, *per*, as in miles per gallon (*mpg*) or feet per second (*ft/sec*).

For example: $\dfrac{20\,miles}{1\,hour} = 20$ *miles per hour*, or 20 *mph*

Average speed is a type of unit rate expressed as distance over time; that is, $r = \dfrac{d}{t}$.

From this formula, we can also derive the formulas for distance, $d = rt$, and for time, $t = \dfrac{d}{r}$.

A **unit price** is a unit rate that contains a price in the numerator (such as *price per pound*).

Model Problem

Samantha drives 250 miles and uses 20 gallons of gasoline. What is her vehicle's gas mileage in miles per gallon (*mpg*)?

Solution:

$$\dfrac{250\,miles}{20\,gallons} = \dfrac{250}{20}\,mpg = 12.5\,mpg$$

Explanation:

Divide $250 \div 20$.

Proportions

A **proportion** is an equation stating that two ratios are equal.

For examples:
$$\frac{3}{2} = \frac{6}{4} \qquad 3:2 = 6:4$$

When a proportion is expressed as equal fractions, **cross-multiplication** yields equal products.

For example: Since $\frac{3}{2} = \frac{6}{4}$, $\quad 2 \times 6 = 3 \times 4$

Model Problem
A computer can perform 360 instructions in 10 microseconds. How many microseconds will it take the computer to perform 288 instructions?

Solution:

(A)   (B)   (C)

$$\frac{360}{10} = \frac{288}{x} \qquad 360x = 2880 \qquad x = 8$$

Explanation of Steps:
  (A) Write a proportion from the given equal ratios or rates. Use a variable, x, for the unknown value. Make sure the numerators *[instructions]* and denominators *[microseconds]* are in the same units.
  (B) Cross-multiply.
  (C) Solve for x.

Percents

Changing to and from percents

A **decimal is changed to percent** by moving the decimal point 2 places to the right and adding a percent sign.

For example: $0.125 = 12.5\%$

A **fraction is changed to percent** by first changing the fraction to a decimal (by dividing on a calculator) and then changing the decimal to percent.

For example: $\dfrac{1}{8} = 1 \div 8 = 0.125 = 12.5\%$

A **percent is changed to a decimal** by moving the decimal point 2 places to the left and removing the percent sign.

For example: $4.5\% = 0.045$

 A **percent is changed to a fraction** by first changing it to a decimal and then using the calculator's $\boxed{\text{MATH}}$ $\boxed{\text{Frac}}$ function to change the decimal to a fraction.

For example: $4.5\% = 0.045$ → $\boxed{\text{MATH}}$ $\boxed{\text{Frac}}$ → 9/200

Rounding percents

If asked to round a percent, the specified place refers to the number when it is written as a percent, not as a decimal.

For examples: Round $\dfrac{3}{8}$ to the *nearest whole percent*. $\dfrac{3}{8} = 0.375 = 37.5\% \approx 38\%$

Round $\dfrac{2}{3}$ to the *nearest tenth of a percent*. $\dfrac{2}{3} = 0.\overline{6} = 66.\overline{6}\% \approx 66.7\%$

Percent Problems

Percent problems can often be solved using simple algebraic proportions. First, state the problem in the following form:

a is p% of b.   *a* is the part, *b* is the whole, and *p* is the percent

Then use the following proportion to solve for the missing value:

$$\frac{a}{b} = \frac{p}{100}$$

A **discount** is a percent that is **subtracted from** a product's original price. A **sales tax** is a percent that is **added to** a product's selling price (*after any discounts are applied*).

Model Problem 1:   What is 40% of 320?

Solution:

$$\frac{a}{320} = \frac{40}{100} \qquad 100a = 12,800 \qquad a = 128$$

Model Problem 2:   15 is 20% of what number?

Solution:

$$\frac{15}{b} = \frac{20}{100} \qquad 1500 = 20b \qquad b = 75$$

Model Problem 3:   12 is what percent of 600?

Solution:

$$\frac{12}{600} = \frac{p}{100} \qquad 1200 = 600p \qquad p = 2 \; (Answer = 2\%)$$

Model Problem 4:
There are 234 men at a convention. This is 36% of the attendees. How many people are attending the convention?

Solution:
234 is 36% of what number?

$$\frac{234}{b} = \frac{36}{100} \qquad 23400 = 36p \qquad p = 650$$

28

Coordinate Graphs

A coordinate plane is a two-dimensional space formed when two number lines, a horizontal **x-axis** and a vertical **y-axis**, are placed perpendicular to each other, intersecting at their respective zeroes. The point where they intersect is called the **origin**.

The axes divide the plane into four quadrants, numbered 1 to 4 starting at the upper right quadrant (where both the *x* and *y* values are positive) and continuing counter-clockwise.

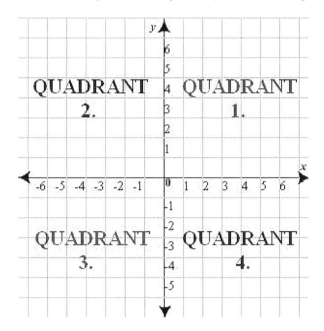

A point on the plane is represented by an **ordered pair**. An ordered pair gives a point's x-coordinate followed by its y-coordinate (always in this order). So, for example, the ordered pair, (3, 2), represents the point on the plane shown below. Use your finger to "travel" along the x-axis to find the x-coordinate, and then use your finger to "travel" up or down, parallel to the y-axis, to find the y-coordinate. The origin is represented by the ordered pair, (0, 0).

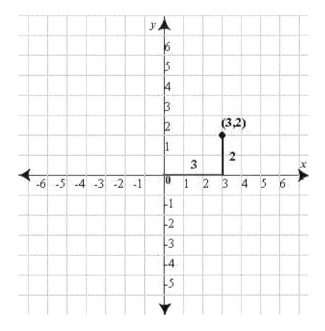

Pythagorean Theorem

In a triangle, the sum of the measures of the three angles is $180°$. In a **right triangle**, one of the angles measures $90°$, and the other two acute angle measures add up to $90°$.

The **legs** of a right triangle are the two shortest sides. The legs are **perpendicular**; they form the right angle. The **hypotenuse** is the longest side.

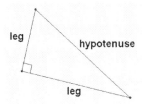

The **Pythagorean Theorem** states a relationship among the sides of any right triangle. If a and b represent the lengths of the legs, and c represents the length of the hypotenuse,

$$a^2 + b^2 = c^2$$

A set of three positive integers that can satisfy this equation is called a **Pythagorean triple**. For examples: {3,4,5}   {5,12,13}   {8,15,17}

Model Problem

A wall is supported by a brace 10 feet long, as shown in the diagram below. If one end of the brace is placed 6 feet from the base of the wall, how many feet up the wall does the brace reach?

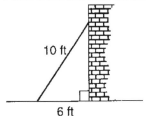

Solution:

(A) $a^2 + b^2 = c^2$

(B) $6^2 + b^2 = 10^2$

(C) $36 + b^2 = 100$

$\qquad b^2 = 64$

$\qquad b = \sqrt{64} = 8\ ft$

Explanation of Steps:

(A) Given two sides of a right triangle, use the Pythagorean Theorem to find the third side.

(B) Substitute given legs as a and b (in either order), and substitute the hypotenuse, if given, as c.
[leg a = 6 and hypotenuse c = 10]

(C) Solve for the remaining variable, and simplify the radical if possible. When taking the square root of both sides, ignore the negative square root since the length of a side must be positive.

II.   LINEAR EQUATIONS AND INEQUALITIES

Properties

Key Terms and Concepts

Identities: The word "identity" comes from the Latin *idem*, meaning the same.
The **additive identity** is zero, because when you add 0 to a value, the value remains the same.
The **multiplicative identity** is one, because when you multiply a value by 1, it remains the same.
    $a + 0 = a$ $a \cdot 1 = a$

Inverses: When an operation is performed between a value and its inverse, the result is the identity for that operation.
So, the **additive inverse** of a number is its opposite, since adding a number and its opposite results in zero, the identity for addition: $a + (-a) = 0$
The **multiplicative inverse** of a number is its reciprocal, since multiplying a number and its reciprocal results in one, the identity for multiplication: $a \cdot \frac{1}{a} = 1$

The **Commutative Property** states that for an operation, the order of the operands doesn't matter. Addition and multiplication are commutative; subtraction and division are not.
In other words, for addition, $a + b = b + a$, and for multiplication, $ab = ba$.

The **Associative Property** states that when performing the same operation on three operands, the order in which we perform the operations (that is, grouping by parentheses) doesn't matter. Addition and multiplication are associative; subtraction and division are not.
In other words, for addition, $a + (b + c) = (a + b) + c$, and for multiplication, $a(bc) = (ab)c$.

The **Distributive Property** states that when a value is multiplied by a sum, we get the same result if we were to multiply the value by each addend separately and then add the products.
In other words, $x(a + b) = xa + xb$.

The **Reflexive Property of Equality** states that a value is equal to itself.
That is, for all real values of a, $a = a$.

The **Symmetric Property of Equality** states that if we switch the two sides of an equation, the equation remains true. That is, if $a = b$, then $b = a$.

Note that if we switch the two sides of an *inequality*, we must also *reverse* the inequality symbol in order for it to remain true. This is called the **Reversal Property of Inequality**.
For examples: If $a > b$, then $b < a$.   If $a \geq b$, then $b \leq a$.

The **Transitive Property of Equality** states that if $a = b$ and $b = c$, then $a = c$.

The **Addition Property of Equality** states that if we add the same value to both sides of an equation, the equality remains. That is, if $a = b$, then $a + c = b + c$.

This extends to the other basic operations as well.

Subtraction Property of Equality: If $a = b$, then $a - c = b - c$.

Multiplication Property of Equality: If $a = b$, then $ac = bc$.

Division Property of Equality: If $a = b$, then $\dfrac{a}{c} = \dfrac{b}{c}$ (where $c \neq 0$).

Closure: A set is **closed** under an operation if, for <u>every pair</u> of elements, when the operation is performed on them, the result is an element of the same set.

For examples: (1) The set of integers is closed under addition because, whenever we add two integers, the result is always another integer.

(2) The set of integers is *not* closed under division because when we divide two integers we *may* get a result that is a fraction (decimal) and not an integer. For example, if we divide 1 by 2, the result is one half, which is not an integer.

The table below shows whether the given sets are closed under the specified operations. Note that division is *not* closed for any set that includes zero, since division by zero is undefined.

	Addition	Subtraction	Multiplication	Division
WHOLE NUMBERS	Y	N	Y	N
INTEGERS	Y	Y	Y	N
RATIONAL NUMBERS	Y	Y	Y	Y*
REAL NUMBERS	Y	Y	Y	Y*

* The sets of <u>non-zero</u> rational or real numbers only.

We can show that the set of **rational numbers is closed under multiplication**:

If $\dfrac{a}{b}$ and $\dfrac{c}{d}$ are rational numbers and a, b, c, and d are integers,

then $\dfrac{a}{b} \times \dfrac{c}{d} = \dfrac{ac}{bd}$ by the rules for multiplying fractions. Since the set of integers is

closed under multiplication, then ac and bd are integers, so $\dfrac{ac}{bd}$ is rational.

Similarly, we can show that the set of **rational numbers is closed under addition**:

If $\dfrac{a}{b}$ and $\dfrac{c}{d}$ are rational numbers and a, b, c, and d are integers, then $\dfrac{a}{b} + \dfrac{c}{d} = \dfrac{ad + bc}{bd}$

by the rules for adding fractions. Since the set of integers is closed under addition and

multiplication, then $ad + bc$ and bd are integers, so $\dfrac{ad + bc}{bd}$ is rational.

Model Problem 1: *understanding the properties of real numbers*

Justify the statement, "Subtraction is *not* associative."

Solution:

If subtraction were associative, then $a-(b-c)$ would equal $(a-b)-c$ for all real numbers a, b, and c. For example, $10-(5-2)$ would equal $(10-5)-2$. However, $10-(5-2)=7$ and $(10-5)-2=3$. So, subtraction is not associative.

Explanation of Steps:

When asked to justify that something is not true, it is usually best to offer a counterexample.

Practice Problems

1. What is the additive inverse of $\frac{2}{3}$?	2. What is the multiplicative inverse of $\frac{2}{3}$?
3. Justify the statement, "Division is *not* commutative."	4. Is the set of whole numbers closed under subtraction? Justify your answer.
5. To find the product of x and 5, which property allows us to perform the following step? $$x \cdot 5 = 5x$$	6. When adding 2 to the sum, $3+x$, which property allows us to perform the first step below? $$2+(3+x)=$$ $$(2+3)+x=$$ $$5+x$$
7. What is the additive inverse of the expression $-ab$?	8. What is the multiplicative inverse of the expression $-\dfrac{1}{ab}$?
9. Is the set of *non-zero* integers closed under division? Justify your answer.	10. Show that the set of *non-zero* rational numbers is closed under division.

Model Problem 2: *application of the distributive property*

Distribute: $-4(1-x)$

Solution: $-4(1-x) = -4(1) + (-4)(-x) = -4 + 4x$

Explanation of steps:

The distributive property, $x(a+b) = xa + xb$, allows us to multiply the value outside the parentheses by each value inside the parentheses, and add the products.

$[So, -4(1-x) = (-4)(1) + (-4)(-x)]$

Practice Problems

11. Distribute: $5(x+5)$	12. Distribute: $4(b-4)$
13. Distribute: $-2(x-1)$	14. Distribute: $-3(a-b)$
15. Simplify: $(1+y)(-1)$	16. Simplify: $-(-a-1)$
17. Apply the distributive property to: $rs + rt$	18. Apply the distributive property to: $2x + 10$

REGENTS QUESTIONS

Multiple Choice

1. The expression $2+3(x+y)$ is equivalent to
   - (1) $5x+5y$
   - (2) $3+2(x+y)$
   - (3) $(2+3)(x+y)$
   - (4) $2+3x+3y$

2. Which is an illustration of the associative property of multiplication?
   - (1) $(ab)c=a(bc)$
   - (2) $a(0)=0$
   - (3) $ab=ba$
   - (4) $a(1)=a$

3. Which statement is *false*?
   - (1) The set of integers is closed under subtraction.
   - (2) The set of integers is closed under division.
   - (3) The set of natural numbers is closed under addition.
   - (4) The set of natural numbers is closed under multiplication.

4. Which property is illustrated by the equation $ax+ay=a(x+y)$?
   - (1) associative
   - (2) commutative
   - (3) distributive
   - (4) identity

5. The statement $2 + 0 = 2$ is an example of the use of which property of real numbers?
   - (1) associative
   - (2) additive identity
   - (3) additive inverse
   - (4) distributive

6. What is the additive inverse of the expression $a - b$?
   - (1) $a + b$
   - (2) $a - b$
   - (3) $-a + b$
   - (4) $-a - b$

7. Which equation illustrates the associative property?
   - (1) $x+y+z=x+y+z$
   - (2) $x(y+z)=xy+xz$
   - (3) $x+y+z=z+y+x$
   - (4) $(x+y)+z=x+(y+z)$

8. Which equation is an example of the use of the associative property of addition?
   - (1) $x+7=7+x$
   - (2) $3(x+y)=3x+3y$
   - (3) $(x+y)+3=x+(y+3)$
   - (4) $3+(x+y)=(x+y)+3$

9. Which statement illustrates the additive identity property?
   - (1) $6+0=6$
   - (2) $-6+6=0$
   - (3) $4(6 + 3) = 4(6) + 4(3)$
   - (4) $(4 + 6) + 3 = 4 + (6 + 3)$

10. The equation $3(4x)=(4x)3$ illustrates which property?
    - (1) commutative
    - (2) associative
    - (3) distributive
    - (4) multiplicative inverse

11. Which equation illustrates the multiplicative inverse property?

    (1) $a \cdot 1 = a$ (3) $a\left(\dfrac{1}{a}\right) = 1$

    (2) $a \cdot 0 = 0$ (4) $(-a)(-a) = a^2$

12. When solving the equation $4(3x^2 + 2) - 9 = 8x^2 + 7$, Emily wrote $4(3x^2 + 2) = 8x^2 + 16$ as her first step. Which property justifies Emily's first step?
        (1) addition property of equality
        (2) commutative property of addition
        (3) multiplication property of equality
        (4) distributive property of multiplication over addition

Constructed Response

13. What is the multiplicative inverse of $3x$?

14. Perform the indicated operation: $-6(a - 7)$
    State the name of the property used.

Solving Linear Equations in One Variable

Key Terms and Concepts

An **equation** is a statement that one expression is equal to another. It contains an = sign.
For example: $3x - 1 = x + 5$

Domain (replacement set): the set of numbers that can replace a variable. Usually, you can assume the domain is {real numbers}. However, if, for example, the variable represents the length of the side of a polygon, the domain should be {positive real numbers}.
Solution set: set of values from the domain that make an equation or inequality true.
Solution: each element of the solution set.

A **variable term** in an equation includes the variable as a factor.
A **constant term** in an equation does not include a variable factor.
For example: $3x + 5 = 35$ has a variable term $[3x]$ and a constant term $[+5]$ on the left side.

The goal when solving an equation is to **isolate the variable** (transform it into $x = a\ value$).

Like Terms: have the same exact variable parts (the numerical coefficients may differ). Like terms may be combined (added or subtracted); unlike terms may not.
For examples: Like terms: $2x$ and $3x$ $-4x^2y$ and x^2y

   Not like terms: $2x$ and $3x^2$ $-4x^2y$ and x^2

To combine like terms:
   (a) Add or subtract the coefficients
   (b) Keep the variable part the same
For example: $2x + 5x - 2y\ \ = 7x - 2y$ the $2x$ and $5x$ are like terms

To solve a linear equation for one variable:
  1. **SIMPLIFY:** Simplify each side of the equation to one or two terms.
     (a) Use the **distributive property** where possible to remove parentheses.
     (b) **Combine like terms** on the same side of the equal sign where possible.
  2. **VARIABLE TERMS TO ONE SIDE:** If there are variable terms on both sides, eliminate a variable term from one side by adding its opposite to both sides.
     Eliminate from which side? Here's a good rule of thumb…
     (a) If one side has more terms, eliminate from that side.
     (b) Otherwise, eliminate the term with the smaller coefficient.
  3. **ISOLATE THE VARIABLE:** Use the reverse order of operations to isolate the variable:
     (c) add the opposite (additive inverse) of the constant term to both sides.
     (d) divide both sides by (or multiply both sides by the reciprocal, or multiplicative inverse, of) the variable term's coefficient.

To check your solution:
Substitute your solution for the variable in the original equation. *It is usually best to use parentheses around the value when substituting.* Then, evaluate both sides of the equation to determine whether the solution makes the equation true.

Checking on the calculator: You can also use the calculator to check your solution.

1. Use the $\boxed{\text{STO►}}$ and $\boxed{\text{ALPHA}}$ buttons to store your answer into the variable.
2. Type in the equation, using $\boxed{\text{2nd}}$ [TEST] to enter the $\boxed{=}$.
3. The calculator will display $\boxed{1}$ If the equation is true for the given value of the variable, or $\boxed{0}$ if it is false.

The Model Problems below will show examples of how to check using the calculator.

Model Problem 1: *equations needing to be simplified first*

Solve: $5(x-2)+3x=30$

Solution:

$$5(x-2)+3x=30$$
(A) $\quad 5x-10+3x=30$
(B) $\qquad 8x-10=30$
(C) $\qquad\qquad 8x=40$
$$\qquad\qquad x=5$$

Explanation of steps:

(A) Distribute *[$5(x-2)=5x-10$]*
(B) Combine like terms [$5x+3x=8x$]
(C) Solve the two-step equation *[add 10 to both sides then divide both sides by 8].*

Check by substituting: $5(5-2)+3(5)=30$
Evaluating the left side gives us $30=30$ ✓

Check on calculator: $5\ \boxed{\text{STO►}}\ \boxed{\text{ALPHA}}\ [\text{X}]\ \boxed{\text{ENTER}}$

($\boxed{1}$ means true.) $5(\ \boxed{\text{ALPHA}}\ [\text{X}] - 2) + 3\ \boxed{\text{ALPHA}}\ [\text{X}]\ \boxed{\text{2nd}}\ [\text{TEST}]\ \boxed{=}\ 30\ \boxed{\text{ENTER}}$

Practice Problems

1. Solve: $\quad 3(m-2)=18$	2. Solve: $\quad 4n-n=-12$
3. Solve: $\quad -5=-(y+1)-y$	4. Solve: $\quad 15x-3(3x+4)=6$

38

<u>Model Problem 2</u>: *equations with variables on both sides*
Solve for x: $2(x-3)-3=8x+3$

Solution:

$$2(x-3)-3=8x+3$$

(A) $2x-6-3=8x+3$

$$2x-9=8x+3$$

(B) $-2x\quad=-2x$

$$-9=6x+3$$

(C) $-12=6x$

$$-2=x$$

Explanation of steps:

(A) Simplify the left side by distributing
[$2(x-3)$ *becomes* $2x-6$ *]* and then by
combining like terms *[* $-6-3$ *becomes* *–9].*

(B) Since there are variables on both sides, eliminate
the variable term with the smaller coefficient
[$2x$*, since* $2<8$ *]* by adding its opposite
[$-2x$ *]* to both sides *[* $8x-2x=6x$ *].*

(C) Now solve the simpler equation *[by adding –3 to
both sides, then dividing both sides by 6].*

Check on calculator: $(-)$ 2 [STO▶] [ALPHA] [X] [ENTER]

($\boxed{1}$ means true.)   2([ALPHA] [X] -3) -3 [2nd] [TEST] $\boxed{=}$ 8 [ALPHA] [X] $+3$ [ENTER]

Practice Problems

5. Solve for x: $3x+8=5x$	6. What is the value of p in the equation $8p+2=4p-10$?
7. What is the value of x in the equation $5(2x-7)=15x-10$?	8. Solve for x: $5(x-2)=2(10+x)$
9. Solve for y: $-4(y-3)=5(2y-6)$	10. Solve for x: $3(x-2)-2(x+1)=5(x-4)$

REGENTS QUESTIONS

Multiple Choice

1. Which value of p is the solution of $5p - 1 = 2p + 20$?

   (1) $\dfrac{19}{7}$ (3) 3

   (2) $\dfrac{19}{3}$ (4) 7

2. Debbie solved the linear equation $3(x + 4) - 2 = 16$ as follows:

[Line 1]	$3(x + 4) - 2 = 16$
[Line 2]	$3(x + 4) = 18$
[Line 3]	$3x + 4 = 18$
[Line 4]	$3x = 14$
[Line 5]	$x = 4\frac{2}{3}$

   She made an error between lines
   (1) 1 and 2   (3) 3 and 4
   (2) 2 and 3   (4) 4 and 5

3. What is the value of x in the equation $2(x - 4) = 4(2x + 1)$?

   (1) –2   (3) $-\dfrac{1}{2}$

   (2) 2   (4) $\dfrac{1}{2}$

4. The value of y in the equation $0.06y + 200 = 0.03y + 350$ is
   (1) 500   (3) 5,000
   (2) $1,666.\overline{6}$ (4) $18,333.\overline{3}$

5. The solution of the equation $5 - 2x = -4x - 7$ is
   (1) 1   (3) –2
   (2) 2   (4) –6

6. When solving for the value of x in the equation $4(x - 1) + 3 = 18$, Aaron wrote the following lines on the board.

$$[\text{line 1}] \qquad 4(x-1)+3=18$$
$$[\text{line 2}] \qquad 4(x-1)=15$$
$$[\text{line 3}] \qquad 4x-1=15$$
$$[\text{line 4}] \qquad 4x=16$$
$$[\text{line 5}] \qquad x=4$$

Which property was used incorrectly when going from line 2 to line 3?
(1) distributive　　　　　　　(3) associative
(2) commutative　　　　　　　(4) multiplicative inverse

7. Which value of x is the solution of the equation $2(x-4)+7=3$?
(1) 1　　　　　　　(3) 6
(2) 2　　　　　　　(4) 0

Constructed Response

8. Solve for y: $5y - 2(y-5) = 19$

9. Solve for g: $3 + 2g = 5g - 9$

10. A method for solving $5(x-2) - 2(x-5) = 9$ is shown below. Identify the property used to obtain each of the two indicated steps.

$$5(x - 2) - 2(x - 5) = 9$$

(1)　$5x - 10 - 2x + 10 = 9$ 　　　(1)　_____

(2)　$5x - 2x - 10 + 10 = 9$ 　　　(2)　_____

$$3x + 0 = 9$$
$$3x = 9$$
$$x = 3$$

11. Solve algebraically for x: $3(x+1) - 5x = 12 - (6x - 7)$

Solving Linear Inequalities in One Variable

Key Terms and Concepts

An **inequality** is a statement that one expression is compared to, but not equal to, another. It contains one of the other relational symbols: $<, >, \leq, \geq,$ or $\neq$.
For example: $y < 2x + 1$

Solving a simple inequality: Follow the same steps as in solving an equation EXCEPT when multiplying or dividing both sides by a <u>negative</u> number, <u>REVERSE</u> the inequality symbol.
For example: If $-3x \leq 15$, then dividing both sides by -3 gives us $x \geq -5$.

It may help to "flip" after solving: Once the variable is isolated, if it appears on the right side of the inequality, it is usually helpful to "flip" the entire inequality (switch the two sides and reverse the inequality symbol) so that the variable appears on the left side instead.
For example: The inequality $6 > y$ can be "flipped" to $y < 6$.

Graphing a simple inequality: Once a simple inequality is solved for a variable, it can be graphed on a number line using the following steps.
  1) On the number line, find the value that the variable is being compared to.
  2) If the inequality symbol is:
     > or <, draw an open circle ($\circ$) at that value, which means it is *not* in the solution set.
     $\geq$ or $\leq$, draw a closed circle ($\bullet$) at that value, which means it is in the solution set.
  3) If the variable is > or $\geq$ the value, shade an arrow to the right; otherwise, if the variable is < or $\leq$ the value, shade an arrow to the left. All the values represented by the shaded arrow, extending infinitely in that direction, are in the solution set.

A **compound inequality chain** has two inequality symbols with the variable in the middle part.
For example: $-4 < x \leq 1$ means $-4 < x$ *and* $x \leq 1$

Solving a compound inequality chain: If the variable is not isolated, perform the inverse operations to eliminate everything around the variable, but do so to *all parts* of the inequality. Be sure to reverse the inequality symbols when multiplying or dividing by a negative number.
For example: Solve $0 < x - 1 < 5$ by adding 1 to all three parts, resulting in $1 < x < 6$.

It may help to "flip" after solving: Once the variable is isolated, if the inequality symbols are > or $\geq$, it is usually helpful to "flip" the entire compound inequality – making sure to reverse the symbols as well – so that the lower limit is on the left and the upper limit is on the right.
For example: $6 > x \geq 2$ can be "flipped" to $2 \leq x < 6$.

To graph a compound inequality chain: Once a compound inequality is solved and arranged so that only < or $\leq$ symbols are used, it can be graphed as follows:
  1) On the number line, find the lower limit and draw an open (for <) or closed (for $\leq$) circle. Then find the upper limit and draw an open (for <) or closed (for $\leq$) circle.
  2) Shade the number line between the lower and upper limits.
For example: $-1 \leq x < 2$ is graphed as

Compound inequality joined by "and": Two simple inequalities joined by "and" can be combined into a compound inequality chain. It may require solving and/or flipping.

For example: $x > 4$ *and* $x \leq 10$ can be combined into the one chain, $4 < x \leq 10$, by flipping the first inequality and then joining them at the x.

Compound inequality joined by "or": Two simple inqualities joined by "or" cannot be combined into a chain. Any solution to *either one or the other* inequality (*or both*) will solve the compound inequality. So, we can graph them on the number line, where they may overlap.

For examples: (1) The compound inequality $x \leq -4$ *or* $x > 3$ can be graphed as shown below.

(2) Consider the compound inequality $x \leq -1$ *or* $x < 3$. Looking at the graphs of the simple inequalities separately, we can easily see that if $x \leq -1$, then certainly $x < 3$. So, the compound inequality is the overlap of the two simple inequalities, or simply $x < 3$.

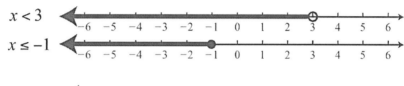

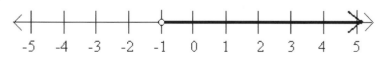

Model Problem 1: *simple inequalities*

Solve and graph the solution set: $5 - 2x < 7$

Solution:

$$7 > 5 - 2x$$

(A) $\underline{\quad -5 \quad -5 \quad}$

$$\frac{2 > -2x}{-2 \quad -2}$$

(B) $-1 < x$

(C) $x > -1$

(D)

Explanation of steps:

(A) Solve like an equation. *[Here, we subtract 5 from both sides, then divide both sides by –2.]*

(B) If, in the process of solving, we multiply or divide both sides by a negative number, reverse the inequality symbol. *[Since we divided by –2, we reverse the symbol from > to <.]*

(C) After solving, if the variable is on the right side, it is helpful to "flip" the entire inequality, including the inequality symbol, so that it appears on the left side.

(D) Graph the result *[open circle, arrow to the greater side]*.

Practice Problems

1. Write an inequality that represents the graph below.  -5 -4 -3 -2 -1 0 1 2 3 4 5	2. Write an inequality that represents the graph below. 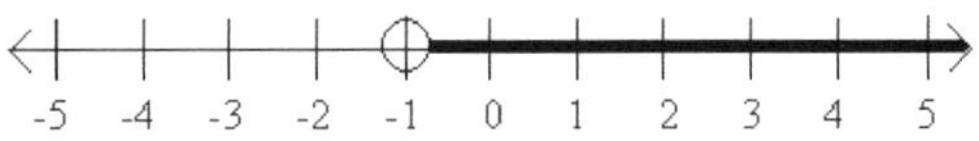 -5 -4 -3 -2 -1 0 1 2 3 4 5
3. Solve and graph the solution set: $2x - 5 \leq 11$	4. Solve and graph the solution set: $-6y + 1 > 25$
5. Solve and graph the solution set: $-4 > 2(r - 3)$	6. Solve and graph the solution set: $-\dfrac{4}{3}(x - 3) \geq 12$
7. Solve: $-4(2m - 6) + m > 3m + 4$	8. Solve: $-5(p + 1) \geq -p + 11$

Model Problem 2: *compound inequalities*

Solve and graph the solution set: $-4 < r - 5 \le -1$

Solution:

$$-4 < r - 5 \le -1$$

(A) $\quad +5 \qquad +5 \quad +5$

$$1 < r \le 4$$

(B)

Explanation of steps:

(A) Solve like any inequality by isolating the variable, but whatever operation is done to the middle part must be done to the other two. The rule still applies: when multiplying or dividing by a negative, reverse the inequality symbols.

(B) Graph the result.

Practice Problems

9. Write a compound inequality represented by the graph below. 	10. Write a compound inequality represented by the graph below. 
11. Write a compound inequality represented by the graph below. 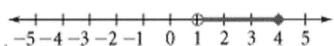	12. Graph $-1 < x < 3$ on the number line. 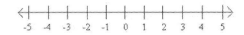
13. Solve and graph the solution set: $\quad 3 \le 2x + 1 < 9$	14. How many positive integers are in the solution set of $-2 < 3x + 4 \le 10$?

REGENTS QUESTIONS

Multiple Choice

1. The inequality $3 + 2x > 5$ is equivalent to
    (1) $x < 1$ (3) $x < 4$
    (2) $x > 1$ (4) $x > 4$

2. Which inequality is shown on the accompanying graph?

    (1) $-4 < x < 2$ (3) $-4 < x \le 2$
    (2) $-4 \le x < 2$ (4) $-4 \le x \le 2$

3. Which inequality is represented in the accompanying graph?

    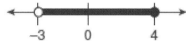

    (1) $-3 \le x < 4$ (3) $-3 < x < 4$
    (2) $-3 \le x \le 4$ (4) $-3 < x \le 4$

4. Which value of x is in the solution set of the inequality $-2x + 5 > 17$?
    (1) -8   (3) -4
    (2) -6   (4) 12

5. Which value of x is in the solution set of the inequality $-4x + 2 > 10$?
    (1) -2   (3) 3
    (2) 2   (4) -4

6. Which value of x is in the solution set of $\dfrac{4}{3}x + 5 < 17$?
    (1) 8   (3) 12
    (2) 9   (4) 16

7. Which value of x is in the solution set of the inequality $-2(x - 5) < 4$?
    (1) 0   (3) 3
    (2) 2   (4) 5

8. What is the solution of $3(2m - 1) \le 4m + 7$?
    (1) $m \le 5$ (3) $m \le 4$
    (2) $m \ge 5$ (4) $m \ge 4$

9. What is the solution of the inequality $-6x - 17 \geq 8x + 25$?
    (1) $x \geq 3$ (3) $x \geq -3$
    (2) $x \leq 3$ (4) $x \leq -3$

10. Which value of x is in the solution set of $-3x + 8 \geq 14$?
    (1) -3 (3) 0
    (2) -1 (4) 3

11. The statement $\left|-15\right| < x < \left|-20\right|$ is true when x is equal to
    (1) -16 (3) 17
    (2) -14 (4) 21

12. Which graph represents the solution set of $2x - 5 < 3$?

    (1)
    (2)
    (3)
    (4)

13. What is the solution of $4x - 30 \geq -3x + 12$?
    (1) $x \geq 6$ (3) $x \geq -6$
    (2) $x \leq 6$ (4) $x \leq -6$

Constructed Response

14. Given: $A = \{18, 6, -3, -12\}$

    Determine all elements of set A that are in the solution of the inequality $\dfrac{2}{3}x + 3 < -2x - 7$

15. Solve algebraically for x: $2(x - 4) \geq \dfrac{1}{2}(5 - 3x)$

16. Solve the inequality $-5(x - 7) < 15$ algebraically for x.

17. Given $2x + ax - 7 > -12$, determine the largest integer value of a when $x = -1$.

Solving Proportions by Linear Equations

Key Terms and Concepts

If two algebraic fractions are equal, the equation represents a proportion. To solve a proportion, we can cross-multiply and then solve the resulting equation.

Model Problem

Solve for x: $\dfrac{3x+1}{5} = \dfrac{2x}{3}$

Solution:

(A) $5(2x) = 3(3x+1)$

(B) $10x = 9x + 3$

         $x = 3$

Explanation of Steps:

(A) Cross-multiply. Be sure to use parentheses where needed.

(B) Solve for *x*.

Practice Problems

1. Solve for *x*: $\dfrac{2}{x+1} = \dfrac{5}{15}$	2. Solve for *p*: $\dfrac{p-5}{4} = \dfrac{p+6}{5}$
3. Solve for *y*: $\dfrac{7y-5}{3} = \dfrac{9y}{4}$	4. Solve for *h*: $\dfrac{1-h}{5} = \dfrac{h-4}{-2}$

REGENTS QUESTIONS

Multiple Choice

1. What is the solution of $\dfrac{k+4}{2} = \dfrac{k+9}{3}$?

        (1) 1   (3) 6

        (2) 5   (4) 14

2. Which value of x is the solution of $\dfrac{2x-3}{x-4} = \dfrac{2}{3}$?

        (1) $-\dfrac{1}{4}$ (3) -4

        (2) $\dfrac{1}{4}$ (4) 4

Constructed Response

3. Solve for y: $\dfrac{y+2}{6} = \dfrac{7}{3}$

4. Solve for n: $n-4 = \dfrac{n-1}{4}$

5. Solve for y: $\dfrac{3+y}{7} = \dfrac{y-9}{3}$

6. Solve for x: $\dfrac{x+2}{5} = \dfrac{x-2}{3}$

7. Solve for x: $\dfrac{x}{6} = \dfrac{x-2}{5}$

8. Solve for P: $\dfrac{3}{4} = \dfrac{P+2}{12}$

Solving Equations with Fractions

Key Terms and Concepts

When an equation involves fractions but more than one term on either side of the equation (unlike last section), we can solve the equation by **eliminating the fractions.** We can accomplish this by **multiplying each term** of the equation by the **least common multiple (LCM)** of the denominators, also called the **least common denominator (LCD).** This will eliminate the denominator of each term since it will divide evenly into the LCD.
(For this course, equations of this type will have no variables in the denominators.)

Note: To learn how to find the LCD, see the *Pre-Algebra Review* section called *GCF and LCM.*

Why can we multiply each term by the LCD? This is because we are actually multiplying both sides of the equation by the LCD (*multiplication property of equality*) and then distributing, without actually showing the distribution step.

For example: To solve $\dfrac{x}{4} - \dfrac{1}{2} = 2$,

$$\boxed{4\left(\frac{x}{4} - \frac{1}{2}\right) = 4(2)} \Rightarrow 4\left(\frac{x}{4}\right) - 4\left(\frac{1}{2}\right) = 4(2) \Rightarrow x - 2 = 8 \Rightarrow x = 10$$

you may skip this step

Model Problem

Solve for x: $\dfrac{x+1}{4} - \dfrac{2x}{3} = \dfrac{x}{12}$

Solution:

(A) $12\left(\dfrac{x+1}{4}\right) - 12\left(\dfrac{2x}{3}\right) = 12\left(\dfrac{x}{12}\right)$

(B) $3(x+1) - 4(2x) = x$

(C) $3x + 3 - 8x = x$

$-5x + 3 = x$

$3 = 6x$

$\frac{1}{2} = x$

Explanation of Steps:

(A) Find the LCD *[12]*. Multiply each term by the LCD.

(B) Each denominator will divide evenly into the LCD, thereby eliminating the denominator.

(C) Solve the resulting equation.

Practice Problems

1. Solve for *x*: $\dfrac{x}{16} + \dfrac{1}{4} = \dfrac{1}{2}$	2. Solve for *x*: $\dfrac{x}{2} + \dfrac{x}{6} = 2$?
3. Solve for x: $\dfrac{3}{5}x + \dfrac{2}{5} = 4$	4. Solve for *x*: $\dfrac{3}{4}x + 2 = \dfrac{5}{4}x - 6$?
5. Solve for x: $\dfrac{3}{4}x = \dfrac{1}{3}x + 5$	6. Solve for x: $\dfrac{3}{4}(x + 3) = 9$
7. Solve for x: $\dfrac{1}{2}(18 - 5x) = \dfrac{1}{3}(6 - 4x)$	8. Solve for x: $\dfrac{2}{3}\left(2x - \dfrac{1}{2}\right) = 13$

REGENTS QUESTIONS

Multiple Choice

1. Solve for x: $x + \dfrac{x-3}{4} = 3$

    (1) -3 (3) 3
    (2) -4 (4) 4

2. Which value of x is the solution of $\dfrac{2x}{5} + \dfrac{1}{3} = \dfrac{7x-2}{15}$?

    (1) $\dfrac{3}{5}$ (3) 3

    (2) $\dfrac{31}{26}$ (4) 7

3. Which value of x is the solution of the equation $\dfrac{2x}{3} + \dfrac{x}{6} = 5$?

    (1) 6 (3) 15
    (2) 10 (4) 30

4. Solve for x: $\dfrac{3}{5}(x+2) = x - 4$

    (1) 8 (3) 15
    (2) 13 (4) 23

5. Which value of x is the solution of $\dfrac{x}{3} + \dfrac{x+1}{2} = x$?

    (1) 1 (3) 3
    (2) -1 (4) -3

6. Which value of x is the solution of the equation $\dfrac{2}{3}x + \dfrac{1}{2} = \dfrac{5}{6}$?

    (1) $\dfrac{1}{2}$ (3) $\dfrac{2}{3}$

    (2) 2 (4) $\dfrac{3}{2}$

7. Which value of x is the solution of the equation $\dfrac{1}{7} + \dfrac{2x}{3} = \dfrac{15x-3}{21}$?

    (1) 6 (3) $\dfrac{4}{13}$

    (2) 0 (4) $\dfrac{6}{29}$

52

8. Which value of x satisfies the equation $\dfrac{7}{3}\left(x+\dfrac{9}{28}\right)=20$?

       (1) 8.25   (3) 19.25

       (2) 8.89   (4) 44.92

Constructed Response

9. Solve for y: $\dfrac{y}{8}-\dfrac{y}{10}=3$

10. Solve for m: $\dfrac{m}{5}+\dfrac{3(m-1)}{2}=2(m-3)$

Literal Equations

Key Terms and Concepts

Literal equations are equations with several variables or letters. To solve a literal equation for a specific variable, perform the same steps to isolate that variable as you would for any equation.

If the specified variable appears in **more than one term** that cannot be combined, use the distributive property to factor it out before dividing both sides by the other factor.

For example: $mx + nx$ can be rewritten as $x(m + n)$ if you need to isolate x.

Model Problem 1: *multi-step literal equations*
Solve for x in terms of a, b, c, and y: $ax + by = c$

Solution:

$$ax + by = c$$

(A)
$$\frac{-by \quad -by}{\;}$$

$$\frac{ax}{a} \quad = \frac{c - by}{a}$$

(B)

$$x = \frac{c - by}{a}$$

Explanation of steps:

(A) Isolate the specified variable *[x]* by first eliminating terms that do not contain the variable. Add the opposite of any such term *[–by]* to both sides.

(B) Then, eliminate any coefficient (numerical or literal) of the specified variable *[a]* by dividing both sides by it.

Practice Problems

1. Solve for p in terms of m: $2m + 2p = 16$	2. Solve for x in terms of b and K: $bx - 2 = K$
3. Solve for m in terms of c and d: $c = 2m + d$	4. Solve for x in terms of a, b, and c: $bx - 3a = c$
5. Solve for w in terms of V, l, and h: $V = lwh$	6. Solve for h in terms of A and b: $A = \dfrac{bh}{2}$

Model Problem 2: *using the distributive property*

Solve for x in terms of a, b, and c: $ax + bx = c$

Solution:

$$ax + bx = c$$

(A) $\quad x(a+b) = c$

(B) $\quad \dfrac{x(a+b)}{a+b} = \dfrac{c}{a+b}$

$$x = \dfrac{c}{a+b}$$

Explanation of steps:

(A) If the variable we're trying to isolate *[x]* appears in more than one term on the same side of the equation, use the distributive property to factor it out.
$$[\,ax + bx = x(a+b)\,]$$

(B) Then, divide both sides by the other factor *[a+b]*.

Practice Problems

7. Solve for x in terms of a and b:	8. Solve for x in terms of a and b:
$3x - ax = b$	$2ax = -bx + 1$

REGENTS QUESTIONS

Multiple Choice

1. If $3ax + b = c,$ then x equals

    (1) $c - b + 3a$

    (2) $c + b - 3a$

    (3) $\dfrac{c - b}{3a}$

    (4) $\dfrac{b - c}{3a}$

2. If the formula for the perimeter of a rectangle is $P = 2l + 2w,$ then w can be expressed as

    (1) $w = \dfrac{2l - P}{2}$

    (2) $w = \dfrac{P - 2l}{2}$

    (3) $w = \dfrac{P - l}{2}$

    (4) $w = \dfrac{P - 2w}{2l}$

3. If $a + ar = b + r,$ the value of a in terms of b and r can be expressed as

    (1) $\dfrac{b}{r} + 1$

    (2) $\dfrac{1 + b}{r}$

    (3) $\dfrac{b + r}{1 + r}$

    (4) $\dfrac{1 + b}{r + b}$

4. The members of the senior class are planning a dance. They use the equation $r = pn$ to determine the total receipts. What is n expressed in terms of r and p?

    (1) $n = r + p$

    (2) $n = r - p$

    (3) $n = \dfrac{p}{r}$

    (4) $n = \dfrac{r}{p}$

5. A formula used for calculating velocity is $v = \dfrac{1}{2} at^2.$ What is a expressed in terms of v and t?

    (1) $a = \dfrac{2v}{t}$

    (2) $a = \dfrac{2v}{t^2}$

    (3) $a = \dfrac{v}{t}$

    (4) $a = \dfrac{v}{2t^2}$

6. If $\dfrac{ey}{n} + k = t$, what is y in terms of e, n, k, and t?

    (1) $y = \dfrac{tn + k}{e}$

    (2) $y = \dfrac{tn - k}{e}$

    (3) $y = \dfrac{n(t + k)}{e}$

    (4) $y = \dfrac{n(t - k)}{e}$

7. If $s = \dfrac{2x+t}{r}$, then x equals

    (1) $\dfrac{rs-t}{2}$

    (2) $\dfrac{rs+1}{2}$

    (3) $2rs-t$

    (4) $rs-2t$

8. If $k = am + 3mx$, the value of m in terms of a, k, and x can be expressed as

    (1) $\dfrac{k}{a+3x}$

    (2) $\dfrac{k-3mx}{a}$

    (3) $\dfrac{k-am}{3x}$

    (4) $\dfrac{k-a}{3x}$

9. The formula for the volume of a pyramid is $V = \dfrac{1}{3}Bh$. What is h expressed in terms of B and V?

    (1) $h = \dfrac{1}{3}VB$

    (2) $h = \dfrac{V}{3B}$

    (3) $h = \dfrac{3V}{B}$

    (4) $h = 3VB$

10. If $rx - st = r$, which expression represents x?

    (1) $\dfrac{r+st}{r}$

    (2) $\dfrac{r}{r+st}$

    (3) $\dfrac{r}{r-st}$

    (4) $\dfrac{r-st}{r}$

11. If $2y + 2w = x$, then w, in terms of x and y, is equal to

    (1) $x-y$

    (2) $\dfrac{x-2y}{2}$

    (3) $x+y$

    (4) $\dfrac{x+2y}{2}$

12. If $abx - 5 = 0$, what is x in terms of a and b?

    (1) $x = \dfrac{5}{ab}$

    (2) $x = -\dfrac{5}{ab}$

    (3) $x = 5 - ab$

    (4) $x = ab - 5$

13. The formula for the volume of a cone is $V = \frac{1}{3}\pi r^2 h$. The radius, r, of the cone may be expressed as

(1) $\sqrt{\dfrac{3V}{\pi h}}$

(3) $3\sqrt{\dfrac{V}{\pi h}}$

(2) $\sqrt{\dfrac{V}{3\pi h}}$

(4) $\dfrac{1}{3}\sqrt{\dfrac{V}{\pi h}}$

Constructed Response

14. Solve for t in terms of A, p, and r: $A = p + prt$

15. Shoe sizes and foot length are related by the formula $S = 3F - 24$, where S represents the shoe size and F represents the length of the foot, in inches.
    a) Solve the formula for F.
    b) To the *nearest tenth of an inch*, how long is the foot of a person who wears a size $10\frac{1}{2}$ shoe?

16. Solve for c in terms of a and b: $bc + ac = ab$

III.   VERBAL PROBLEMS

Translating Expressions

Key Terms and Concepts

To translate verbal expressions into algebraic expressions, look for phrases commonly used to represent operations:

Addition	**Subtraction**	**Multiplication**	**Division**
increased by	decreased by	multiplied by	divided by
sum	difference	product	quotient
plus	minus	times	
more than	*less than*		
added to	*subtracted from*		

Of course, this is not a comprehensive list. For example, just as we can express subtraction using the word, "decreased," we can also use a synonym, such as "diminished" or "reduced."

Order of operands: Generally, operands are written in the order they appear in the verbal expression, with the important exception of the phrases written in *italics* above.

For example: The following are written as $a - b$,   but the following are written as $b - a$

> a decreased by b a less than b
> difference of a and b a subtracted from b
> a minus b

Addition and multiplication are commutative, so the order of their operands shouldn't matter.

Note also: When we multiply a variable by a number, we generally write the number (called the numerical coefficient) first.

For example: "x times 5" is usually written $5x$ rather than $x \cdot 5$

Other common phrases:

Twice	means two times:	"twice x" is written $2x$
Fraction of	means fraction times:	"two-fifths of x" is written $\dfrac{2}{5}x$
The quantity	means parentheses:	"twice the quantity $x + 5$" is written $2(x + 5)$

The placement of **commas** may be important at times.

For example: "Product of x and y, decreased by 2"   means $xy - 2$
                                      "Product of x and y decreased by 2"   means $x(y - 2)$

Writing an expression "in terms of" a variable:
- Often in verbal expressions, a quantity is described by comparing it to another quantity.
- This latter quantity, to which it is being compared, will be represented by a variable.
- This variable quantity will often appear at the end of a verbal clause, and commonly after comparative words such as "than" or "as many as."
- Once the variable quantity is established, then the other quantities will be written as expressions containing, or in terms of, this variable.

For example: If "John's age is 5 less *than* Tom's age," let the variable t represent Tom's age. John's age is represented by an expression in terms of t, namely: $t-5$.

Ratios: if a ratio is stated between two (or more) quantities, then use a variable to represent their common factor.

For example: If the numbers of boys and girls in a class are in the ratio 3:2, then express the number of boys as $3x$ and the number of girls as $2x$.

Consecutive integers are such that each integer is 1 larger than the previous.

For examples: 4, 5, and 6 are three consecutive integers, as are –2, –1, and 0.

If x is the smallest of three consecutive integers, then the three numbers can be expressed as x, $x+1$, and $x+2$. The next consecutive integer after $x+2$ is $x+3$.

Consecutive even integers (for example, 6, 8, 10, and 12) and **consecutive odd integers** (for example, 5, 7, 9, and 11) are each 2 larger than the previous.

So, If x is the smallest of four consecutive even integers, then the four numbers can be expressed as x, $x+2$, $x+4$, and $x+6$.

(The same expressions are used if x is the smallest of four consecutive *odd* integers.)

Monetary values: to find the value of coins or bills, multiply the number of each coin or bill by its denomination, and add the products.

For example: x dimes and $3x$ nickels have a total value of $(0.10)x+(0.05)3x$, or $0.25x$.

Total costs: to find the total cost of items purchased, multiply the number of each item by its price, and add the products.

For example: x apples at \$0.25 each and $(10-x)$ oranges at \$0.40 each costs a total of $(0.25)x+(0.40)(10-x)$, or after simplifying, $4.00-0.15x$.

Chain comparisons: In some cases, a quantity is written as an expression in terms of a variable, and then another quantity is written in terms of *this expression*. Use a variable for the quantity at the *end of the chain*, and then work backwards to express the others.

For example: Mark is 4 years younger than Samuel and Charles is three times as old as Mark. What is the sum of their ages written as an expression?
Mark is described in terms of Samuel and Charles is described in terms of Mark, so the chain is: Charles → Mark → Samuel.   Let s be Samuel's age.
If s is Samuel's age, Mark's age is $s-4$, and Charles' age is $3(s-4)$.
The sum of their ages is $s+(s-4)+3(s-4)$, which simplifies to $5s-16$.

Model Problem 1: *expressions in terms of one variable*

Abby, Barbie, and Carol are sisters. Abby's age is four times Carol's age and Barbie's age if 5 less than twice Carol's age. Write an expression, in terms of Carol's age, c, for the sum of their ages.

Solution:

Carol's age is c.   Abby's age is $4c$.   Barbie's age is $2c-5$.

So the sum of their ages can be represented by $c+4c+2c-5$, which simplifies to $7c-5$.

Explanation of Steps:

A common temptation is to create a separate variable for each quantity *[the three ages]*. But, since all other quantities *[Abby's and Barbie's ages]* are expressed using comparisons to a single value *[Carol's age]*, it is much better to use only one variable *[c]* and then write expressions for the others in terms of this variable.

Practice Problems

1. Write an expression for 5 less than the product of 7 and x.	2. Write an expression for twice the difference of x and 8.
3. The sum of Scott's age and Greg's age is 33 years. If Greg's age is represented by g, what is Scott's age in terms of g?	4. Tara buys two items that cost d dollars each. She gives the cashier $20, which is more than the total cost. Write an expression to represent the change she should receive.
5. John is four times as old as Ashley. If x represents Ashley's age, write an expression to represent how old John will be in 10 years.	6. If n represents the height of an object in inches, write an expression in terms of n to represent the height of the object in feet. (12 inches = 1 foot.)

Model Problem 2: *consecutive integers*

Write an expression, in simplest terms, for the sum of three consecutive odd integers.

Solution:

$x + x + 2 + x + 4 = 3x + 6$

Explanation of Steps:

Let a variable *[x]* represent the smallest of the integers. Write expressions involving this variable for each of the other integers *[x+2 and x+4]*. Express the sum and simplify the result by combining like terms.

Practice Problems

7. If *y* is the smallest of four consecutive integers, write an expression, in simplest terms, for the sum of the four integers.	8. If the smallest of three consecutive even integers is $x + 3$, write the sum of the three integers in simplest terms.

Model Problem 3: *chain comparisons*

Ashanti and Maria went to the store to buy snacks for their back-to-school party. They bought bags of chips, pretzels, and nachos. They bought three times as many bags of pretzels as bags of chips, and two fewer bags of nachos than bags of pretzels. If *x* represents the number of bags of chips they bought, express, in terms of *x*, how many bags of snacks they bought in all.

Solution:

 (A) Let x be the number of bags of chips bought.

 (B) Then, 3x represents the bags of pretzels, and 3x − 2 represents the bags of nachos.

 (C) Altogether, they bought $x+3x+3x-2$, or $7x-2$, bags of snacks.

Explanation of Steps:

 (A) Create the comparison chain and let the variable represent the quantity at the end of the chain. *[Pretzels are expressed in terms of chips and nachos are expressed in terms of pretzels. So, the chain is Nachos → Pretzels → Chips. Let x represent the chips.]*

 (B) By working backwards in the chain, express the other quantities in terms of this variable.

 (C) Combine the expressions using the operation specified.

Practice Problems

9. Camille is 7 years older than Donny, and Donny is 4 years younger than Tommy. Write an expression for the total ages of the three people, in simplest form.	10. The life span of a whale is 4 times than of a stork, which lives 70 years longer than a horse. Write an expression, in simplest form, for the total life spans of the three creatures.

REGENTS QUESTIONS

Multiple Choice

1. The product of two numbers is p. If one number is represented by n, then the other number can be represented by

    (1) $p-n$

    (3) $\dfrac{n}{p}$

    (2) pn

    (4) $\dfrac{p}{n}$

2. If $2x+7$ represents an odd number, what is the next larger odd number?
    (1) $3x+7$ (3) $2x+9$
    (2) $2x+8$ (4) $3x+9$

3. Fifteen postage stamps are purchased; some are 8-cent stamps and the rest are 10-cent stamps. If x represents the number of 8-cent stamps, which expression represents the total cost of the 15 stamps?
    (1) $x+(15-x)$ (3) $.08x+.10(x-15)$
    (2) $.08x+.10(15-x)$ (4) $x+(x-15)$

4. Which verbal expression represents $2(n-6)$?
    (1) two times n minus six   (3) two times the quantity n less than six
    (2) two times six minus n (4) two times the quantity six less than n

5. Mr. Turner bought x boxes of pencils. Each box holds 25 pencils. He left 3 boxes of pencils at home and took the rest to school. Which expression represents the total number of pencils he took to school?
    (1) $22x$ (3) $25-3x$
    (2) $25x-3$ (4) $25x-75$

6. Which verbal expression is represented by $\dfrac{1}{2}(n-3)$?

    (1) one-half n decreased by 3   (3) the difference of one-half n and 3
    (2) one-half n subtracted from 3   (4) one-half the difference of n and 3

7. What is the perimeter of a regular pentagon with a side whose length is $x+4$?
    (1) x^2+16 (3) $5x+4$
    (2) $4x+16$ (4) $5x+20$

8. Tim ate four more cookies than Alice. Bob ate twice as many cookies as Tim. If x represents the number of cookies Alice ate, which expression represents the number of cookies Bob ate?
    (1) $2+(x+4)$ (3) $2(x+4)$
    (2) $2x+4$ (4) $4(x+2)$

9. Which verbal expression can be represented by $2(x-5)$?

    (1) 5 less than 2 times x (3) twice the difference of x and 5

    (2) 2 multiplied by x less than 5   (4) the product of 2 and x, decreased by 5

10. Which algebraic expression represents 15 less than x divided by 9?

    (1) $\dfrac{x}{9}-15$ (3) $15-\dfrac{x}{9}$

    (2) $9x-15$ (4) $15-9x$

11. Timmy bought a skateboard and two helmets for a total of d dollars. If each helmet cost h dollars, the cost of the skateboard could be represented by

    (1) $2dh$ (3) $d-2h$

    (2) $\dfrac{dh}{2}$ (4) $d-\dfrac{h}{2}$

12. Marcy determined that her father's age is four less than three times her age. If x represents Marcy's age, which expression represents her father's age?

    (1) $3x-4$ (3) $4x-3$

    (2) $3(x-4)$ (4) $4-3x$

13. A correct translation of "six less than twice the value of x" is

    (1) $2x<6$ (3) $6<2x$

    (2) $2x-6$ (4) $6-2x$

14. If Angelina's weekly allowance is d dollars, which expression represents her allowance, in dollars, for x weeks?

    (1) dx (3) $x+7d$

    (2) $7dx$ (4) $\dfrac{d}{x}$

15. Which verbal expression is represented by $2(x+4)$?

    (1) twice the sum of a number and four

    (2) the sum of two times a number and four

    (3) two times the difference of a number and four

    (4) twice the product of a number and four

16. Which expression represents "5 less than twice x"?

    (1) $2x-5$ (3) $2(5-x)$

    (2) $5-2x$ (4) $2(x-5)$

17. Which expression represents the number of hours in w weeks and d days?

    (1) $7w+12d$ (3) $168w+24d$

    (2) $84w+24d$ (4) $168w+60d$

18. Julie has three children whose ages are consecutive odd integers. If x represents the youngest child's age, which expression represents the sum of her children's ages?

        (1) $3x+3$ (3) $3x+5$

        (2) $3x+4$ (4) $3x+6$

Constructed Response

19. Write an algebraic expression to represent the total number of cents in q quarters and d dimes.

20. Mr. Cash bought d dollars worth of stock. During the first year, the value of the stock tripled. The next year, the value of the stock decreased by \$1200.

    a) Write an expression in terms of d to represent the value of the stock after two years.

    b) If an initial investment is \$1,000, determine its value at the end of 2 years.

21. Ashanti and Maria went to the store to buy snacks for their back-to-school party. They bought bags of chips, pretzels, and nachos. They bought three times as many bags of pretzels as bags of chips, and two fewer bags of nachos than bags of pretzels. If x represents the number of bags of chips they bought, express, in terms of x, how many bags of snacks they bought in all.

Translating "Each"

Key Terms and Concepts

We have already seen the word "each" used in a number of verbal expressions. Phrases such as "for each," "for every," or "per" are frequently used to represent the **variable term** in a linear expression. Often, the situation will also involve a "starting" or "one-time" value, which usually represents a **constant term**.

For example: Taxis in a certain city charge $2.50 just to enter the taxi and $1.50 <u>for each</u> mile driven. If we let m represent the number of miles, the cost is $2.50 + 1.50m$.

Sometimes the word "per" is omitted but understood.

For example: "$5 a month" means the same as "$5 per month" or "$5 for each month."

Model Problem

Oberon Cell Phone Company advertises service for 3 cents per minute plus a monthly fee of $29.95. Write an expression for the monthly cost if n call minutes are used.

Solution:

$$0.03n + 29.95$$

Explanation:

The monthly fee is a one-time starting cost for the month, so it is the constant term. The cost "per" minute *[3 cents times n]* is the variable term.

Practice Problems

1. Essence of Yoga charges $80 per month with a $75 registration fee. Write an expression for the cost of an x-month membership.	2. Andy deposits $100 in a bank account that earns $5 interest annually. Write an expression for the balance in the account after y years.
3. Abbey starts with $20 and plays an arcade game that costs 50 cents per game. Write an expression for the amount of money remaining after g games are played.	4. An airplane 30,000 feet above the ground begins descending at the constant rate of 2000 feet per minute. Write an expression for the plane's altitude after m minutes.

REGENTS QUESTIONS

Multiple Choice

1. Marie currently has a collection of 58 stamps. If she buys *s* stamps each week for *w* weeks, which expression represents the total number of stamps she will have?

    (1) $58sw$ (3) $58s + w$

    (2) $58 + sw$ (4) $58 + s + w$

2. A company that manufactures radios first pays a start-up cost, and then spends a certain amount of money to manufacture each radio. If the cost of manufacturing *r* radios is given by the function $c(r) = 5.25r + 125$, then the value 5.25 best represents

    (1) the start-up cost
    (2) the profit earned from the sale of one radio
    (3) the amount spent to manufacture each radio
    (4) the average number of radios manufactured

3. A cell phone company charges $60.00 a month for up to 1 gigabyte of data. The cost of additional data is $0.05 per megabyte. If *d* represents the number of additional megabytes used and *c* represents the total charges at the end of the month, which linear equation can be used to determine a user's monthly bill?

    (1) $c = 60 - 0.05d$ (3) $c = 60d - 0.05$

    (2) $c = 60.05d$ (4) $c = 60 + 0.05d$

Translating Equations

Key Terms and Concepts

An **equation** is simply a statement in which two expressions are set equal to each other. So, to translate a verbal problem into an equation:
1. translate each quantity into an expression in terms of a variable (see the last section)
2. determine from the problem how the expressions form an equation

For examples: the sum of x and $2x$ is 42   becomes $x + 2x = 42$

                 $4s$ is 15 more than s becomes $4s = s + 15$

Sometimes a **known formula** needs to be applied in order to write the appropriate equation.
For example: If a rectangle's length is 4x, its width is x, and its area is 100,

            use the formula $A = lw$ to produce the equation, $100 = (4x)(x)$, or $100 = 4x^2$

Model Problem
Write an equation to find three consecutive even integers whose sum is 84.

Solution:
    (A) Let x be the first even integer.
    (B) Therefore, $x + 2$ is the second even integer and $x + 4$ is the third even integer.
    (C) So, $x + (x + 2) + (x + 4) = 84$, or, after simplifying the left side of the equation,
         $3x + 6 = 84$.

Explanation of Steps:
    (A) Determine which quantity will be represented by the variable *[in a consecutive integers problem, x can represent the first in the sequence]*.
    (B) Express the other quantities as expressions in terms of this variable *[each consecutive even integer is two larger than the previous]*.
    (C) Arrange the expressions to form a correct equation *[the sum is 84]*.

Practice Problems

1. The radius of a circle is represented by $3x + 2$, and the length of the diameter is 22 centimeters. Write an equation to find the value of x, in centimeters.	2. Jerome purchased four more apples than oranges. Apples cost 30 cents each and oranges cost 50 cents each. Jerome spent a total of $3.60. Write an equation to find how many oranges he purchased.
3. During a recent winter, the ratio of deer to foxes was 7 to 3 in one county of New York State. If there were 210 foxes in the county, what was the number of deer in the county?	4. There are 357 seniors in Harris High School. The ratio of boys to girls is 7:10. How many boys are in the senior class?

REGENTS QUESTIONS

Multiple Choice

1. When Albert flips open his mathematics textbook, he notices that the product of the page numbers of the two facing pages that he sees is 156. Which equation could be used to find the page numbers that Albert is looking at?
   - (1) $x + (x+1) = 156$
   - (2) $(x+1) + (x+2) = 156$
   - (3) $(x+1)(x+3) = 156$
   - (4) $x(x+1) = 156$

2. Rhonda has $1.35 in nickels and dimes in her pocket. If she has six more dimes than nickels, which equation can be used to determine x, the number of nickels she has?
   - (1) $0.05(x+6) + 0.10x = 1.35$
   - (2) $0.05x + 0.10(x+6) = 1.35$
   - (3) $0.05 + 0.10(6x) = 1.35$
   - (4) $0.15(x+6) = 1.35$

3. If h represents a number, which equation is a correct translation of "Sixty more than 9 times a number is 375"?
   - (1) $9h = 375$
   - (2) $9h + 60 = 375$
   - (3) $9h - 60 = 375$
   - (4) $60h + 9 = 375$

4. The width of a rectangle is 3 less than twice the length, x. If the area of the rectangle is 43 square feet, which equation can be used to find the length, in feet?
   - (1) $2x(x-3) = 43$
   - (2) $x(3-2x) = 43$
   - (3) $2x + 2(2x-3) = 43$
   - (4) $x(2x-3) = 43$

5. If n is an odd integer, which equation can be used to find three consecutive odd integers whose sum is -3?
   - (1) $n + (n+1) + (n+3) = -3$
   - (2) $n + (n+1) + (n+2) = -3$
   - (3) $n + (n+2) + (n+4) = -3$
   - (4) $n + (n+2) + (n+3) = -3$

6. The width of a rectangle is 4 less than half the length. If l represents the length, which equation could be used to find the width, w?
   - (1) $w = \frac{1}{2}(4-l)$
   - (2) $w = \frac{1}{2}(l-4)$
   - (3) $w = \frac{1}{2}l - 4$
   - (4) $w = 4 - \frac{1}{2}l$

7. Three times the sum of a number and four is equal to five times the number, decreased by two. If x represents the number, which equation is a correct translation of the statement?
   - (1) $3(x+4) = 5x - 2$
   - (2) $3(x+4) = 5(x-2)$
   - (3) $3x + 4 = 5x - 2$
   - (4) $3x + 4 = 5(x-2)$

8. The length of the shortest side of a right triangle is 8 inches. The lengths of the other two sides are represented by consecutive odd integers. Which equation could be used to find the lengths of the other sides of the triangle?

    (1) $8^2 + (x+1)^2 = x^2$　　　　　　　(3) $8^2 + (x+2)^2 = x^2$

    (2) $x^2 + 8^2 = (x+1)^2$　　　　　　　(4) $x^2 + 8^2 = (x+2)^2$

9. John has four more nickels than dimes in his pocket, for a total of $1.25. Which equation could be used to determine the number of dimes, x, in his pocket?

    (1) $0.10(x+4) + 0.05(x) = \$1.25$　　　(3) $0.10(4x) + 0.05(x) = \$1.25$

    (2) $0.05(x+4) + 0.10(x) = \$1.25$　　　(4) $0.05(4x) + 0.10(x) = \$1.25$

Translating Inequalities

Key Terms and Concepts

Translate a verbal problem into an **inequality** the same way as you do for an equation, but instead of using an = sign for equality, use the appropriate **inequality symbol**:

$>$	"is more than"	"is greater than"	
$<$	"is less than"		
$\geq$	"greater than or equal to"	"at least"	"not less than"
$\leq$	"less than or equal to"	"at most"	"not more than"

Caution: recognize the distinction in verbal problems between the phrases "more than" and "less than" for addition and subtraction and the phrases "is more than" and "is less than" for an inequality.

For example: 5 more than x $\rightarrow$ $x+5$ 5 is more than x $\rightarrow$ $5>x$

                    5 less than x $\rightarrow$ $x-5$ 5 is less than x $\rightarrow$ $5<x$

Model Problem

Barbara has $10 to buy milk and cookies for her child's party. If each pint of milk costs $0.55 and each cookie costs $0.15, and she wants to buy 6 times as many cookies as pints of milk, write an inequality that shows how much she can buy for at most $10.

Solution:

(A) Let m = the number of pints of milk she buys; $6m$ = the number of cookies she buys.

(B) $0.55m+0.15(6m) \leq 10.00$

Explanation of Steps:

(A) Write expressions in terms of a variable.

(B) Create an appropriate inequality using the expressions.

Practice Problems

1. Allison is nine inches taller than Ben. The sum of their heights is less than 144 inches. Write an appropriate inequality to describe this situation.	2. You need to purchase a apples and b bananas, and you can spend no more than $100. Apples cost 75 cents each and bananas cost one dollar each. Write an inequality, in terms of a and b, to describe this situation.

REGENTS QUESTIONS

Multiple Choice

1. An electronics store sells DVD players and cordless telephones. The store makes a $75 profit on the sale of each DVD player (*d*) and a $30 profit on the sale of each cordless telephone (*c*). The store wants to make a profit of at least $255.00 from its sales of DVD players and cordless phones. Which inequality describes this situation?

    (1) $75d + 30c < 255$ (3) $75d + 30c > 255$

    (2) $75d + 30c \leq 255$ (4) $75d + 30c \geq 255$

2. Students in a ninth grade class measured their heights, *h*, in centimeters. The height of the shortest student was 155 cm, and the height of the tallest student was 190 cm. Which inequality represents the range of heights?

    (1) $155 < h < 190$ (3) $h \geq 155$ or $h \leq 190$

    (2) $155 \leq h \leq 190$ (4) $h > 155$ or $h < 190$

3. Mrs. Smith wrote "Eight less than three times a number is greater than fifteen" on the board. If *x* represents the number, which inequality is a correct translation of this statement?

    (1) $3x - 8 > 15$ (3) $8 - 3x > 15$

    (2) $3x - 8 < 15$ (4) $8 - 3x < 15$

4. The sign shown below is posted in front of a roller coaster ride at the Wadsworth County Fairgrounds.

    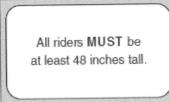

    If *h* represents the height of a rider in inches, what is a correct translation of the statement on this sign?

    (1) $h < 48$ (3) $h \leq 48$

    (2) $h > 48$ (4) $h \geq 48$

5. Roger is having a picnic for 78 guests. He plans to serve each guest at least one hot dog. If each package, *p*, contains eight hot dogs, which inequality could be used to determine how many packages of hot dogs Roger will need to buy?

    (1) $p \geq 78$ (3) $8 + p \geq 78$

    (2) $8p \geq 78$ (4) $78 - p \geq 8$

6. The ninth grade class at a local high school needs to purchase a park permit for $250.00 for their upcoming class picnic. Each ninth grader attending the picnic pays $0.75. Each guest pays $1.25. If 200 ninth graders attend the picnic, which inequality can be used to determine the number of guests, x, needed to cover the cost of the permit?

   (1) $0.75x - (1.25)(200) \geq 250.00$ (3) $(0.75)(200) - 1.25x \geq 250.00$

   (2) $0.75x + (1.25)(200) \geq 250.00$ (4) $(0.75)(200) + 1.25x \geq 250.00$

7. The length of a rectangle is 15 and its width is w. The perimeter of the rectangle is, *at most*, 50. Which inequality can be used to find the longest possible width?

   (1) $30 + 2w < 50$ (3) $30 + 2w > 50$

   (2) $30 + 2w \leq 50$ (4) $30 + 2w \geq 50$

8. Carol plans to sell twice as many magazine subscriptions as Jennifer. If Carol and Jennifer need to sell at least 90 subscriptions in all, which inequality could be used to determine how many subscriptions, x, Jennifer needs to sell?

   (1) $x \geq 45$ (3) $2x - x \geq 90$

   (2) $2x \geq 90$ (4) $2x + x \geq 90$

9. Jeremy is hosting a Halloween party for 80 children. He will give each child *at least* one candy bar. If each bag of candy contains 18 candy bars, which inequality can be used to determine how many bags, c, Jeremy will need to buy?

   (1) $18c \geq 80$ (3) $\dfrac{c}{18} \geq 80$

   (2) $18c \leq 80$ (4) $\dfrac{c}{18} \leq 80$

Word Problems – Linear Equations

Key Terms and Concepts

To solve a word problem using an equation:
1. Translate each quantity into an expression in terms of a variable
2. Determine from the problem how the expressions form an equation
3. Write the equation
4. Solve for the variable, and check your solution
5. Be sure to answer the question asked by the problem, and make sure your answer is reasonable!

How to check your solution: Substitute the solution for the variable into the original equation and evaluate both sides of the equation to make sure the solution makes the equation true.

When is an answer unreasonable?
For examples, if your answer is that the length of a side of a triangle is –4 inches, or that the number of people on a bus is 10.75, dismiss these as unreasonable and retrace your steps.

Model Problem

Tamara has five less than four times as many friendship bracelets as Allison. If they have a total of 55 bracelets, how many bracelets does Tamara have?

Solution:

(A) Let a = how many bracelets Allison has;

(B) Tamara has $4a - 5$ bracelets.

$$a + 4a - 5 = 55$$

(C)
$$5a - 5 = 55$$
$$5a = 60$$
$$a = 12$$

(D) To check the solution:
$$a + 4a - 5 = 55$$
$$(12) + 4(12) - 5 = 55$$
$$12 + 48 - 5 = 55$$
$$55 = 55 \checkmark$$

(E) If Allison has 12 bracelets, then Tamara has
$4a - 5 = 4(12) - 5 = 43$ bracelets.
This is reasonable: $12 + 43 = 55$.
Tamara has 43 bracelets.

Explanation of Steps:

(A) Determine what the variable is.

(B) Express other quantities *[Tamara's bracelets]* in terms of the variable, and write an appropriate equation *[the sum of a and 4a – 5 is 55]*.

(C) Solve for the variable.

(D) Check the solution *[a = 12]* by substituting it into the original equation *[substitute (12) for a]* and then evaluate each side to see if the equation is true.

(E) Answer the question asked by the problem; sometimes the value of the variable does ***not*** answer the question *[a is the number that Allison has, but we need to say how many Tamara has, which is 4a – 5].* Check that the answer is reasonable.

Practice Problems

1. Jamie is 5 years older than her sister Amy. If the sum of their ages is 19, how old is Jamie?	2. Arielle has a collection of grasshoppers and crickets. She has 561 insects in all. The number of grasshoppers is twice the number of crickets. Find the number of *each* type of insect that she has.
3. Three times as many robins as cardinals visited a bird feeder. If a total of 20 robins and cardinals visited the feeder, how many were robins?	4. On the JV baseball team, the number of sophomores is four more than twice the number of freshmen. If there are 16 players combined on the team, how many of each grade level are there?
5. Keisha has 28 video discs, which is 8 less than 4 times the number of video discs in Minnie's collection. How many video discs does Minnie own?	6. There were 100 more balcony tickets than main-floor tickets sold for a concert. The balcony tickets sold for $4 and the main-floor tickets sold for $12. The total amount of sales for both types of tickets was $3,056. Find the number of balcony tickets that were sold.
7. In his piggy bank, Neil has three times as many dimes as nickels and he has four more quarters than nickels. The coins total $4.60. How many of each coin does he have?	8. A craft shop sold 150 pillows. Small pillows were $6.50 each and large pillows were $9.00 each. If the total amount collected from the sale of these items was $1180.00, what is the total number of each size pillow that was sold?

REGENTS QUESTIONS

Multiple Choice

1. At the beginning of her mathematics class, Mrs. Reno gives a warm-up problem. She says, "I am thinking of a number such that 6 less than the product of 7 and this number is 85." Which number is she thinking of?

      (1) 11   (3) 84

      (2) 13   (4) 637

2. Mario paid $44.25 in taxi fare from the hotel to the airport. The cab charged $2.25 for the first mile plus $3.50 for each additional mile. How many miles was it from the hotel to the airport?

      (1) 10   (3) 12

      (2) 11   (4) 13

3. Robin spent $17 at an amusement park for admission and rides. If she paid $5 for admission, and rides cost $3 each, what is the total number of rides that she went on?

      (1) 12   (3) 9

      (2) 2   (4) 4

4. Pam is playing with red and black marbles. The number of red marbles she has is three more than twice the number of black marbles she has. She has 42 marbles in all. How many red marbles does Pam have?

      (1) 13   (3) 29

      (2) 15   (4) 33

5. Sam and Odel have been selling frozen pizzas for a class fundraiser. Sam has sold half as many pizzas as Odel. Together they have sold a total of 126 pizzas. How many pizzas did Sam sell?

      (1) 21   (3) 63

      (2) 42   (4) 84

6. At Genesee High School, the sophomore class has 60 more students than the freshman class. The junior class has 50 fewer students than twice the students in the freshman class. The senior class is three times as large as the freshman class. If there are a total of 1,424 students at Genesee High School, how many students are in the freshman class?

      (1) 202   (3) 235

      (2) 205   (4) 236

7. The ages of three brothers are consecutive even integers. Three times the age of the youngest brother exceeds the oldest brother's age by 48 years. What is the age of the youngest brother?

      (1) 14   (3) 22

      (2) 18   (4) 26

8. Josh and Mae work at a concession stand. They each earn $8 per hour. Josh worked three hours more than Mae. If Josh and Mae earned a total of $120, how many hours did Josh work?

      (1) 6   (3) 12

      (2) 9   (4) 15

9. Michael is 25 years younger than his father. The sum of their ages is 53. What is Michael's age?

      (1) 14   (3) 28

      (2) 25   (4) 39

10. Ben has four more than twice as many CDs as Jake. If they have a total of 31 CDs, how many CDs does Jake have?

      (1) 9   (3) 14

      (2) 13   (4) 22

11. The total score in a football game was 72 points. The winning team scored 12 points more than the losing team. How many points did the winning team score?

      (1) 30   (3) 54

      (2) 42   (4) 60

Constructed Response

12. Two numbers are in the ratio of 1:4 and their sum is 50. What is the *smaller* number?

13. A postal clerk sold 50 postage stamps for $7.00. Some were 2-cent stamps and the rest were 22-cent stamps. Find the number of *each* kind of stamp that was sold.

14. Sara's telephone service costs $21 per month plus $0.25 for each local call, and long-distance calls are extra. Last month, Sara's bill was $36.64, and it included $6.14 in long-distance charges. How many local calls did she make?

15. The sum of the ages of the three Romano brothers is 63. If their ages can be represented as consecutive integers, what is the age of the middle brother?

16. Every month, Omar buys pizzas to serve at a party for his friends. In May, he bought three more than twice the number of pizzas he bought in April. If Omar bought 15 pizzas in May, how many pizzas did he buy in April?

17. The sum of three consecutive odd integers is 18 less than five times the middle number. Find the three integers.

18. The difference between two numbers is 28. The larger number is 8 less than twice the smaller number. Find *both* numbers.

19. Donna wants to make trail mix made up of almonds, walnuts and raisins. She wants to mix one part almonds, two parts walnuts, and three parts raisins. Almonds cost $12 per pound, walnuts cost $9 per pound, and raisins cost $5 per pound. Donna has $15 to spend on the trail mix. Determine how many pounds of trail mix she can make.

Word Problems – Inequalities

Key Terms and Concepts

To solve a word problem using an inequality:
1. Translate each quantity into an expression in terms of a variable
2. Determine from the problem how the expressions form an inequality
3. Write and solve the inequality
4. Be sure to answer the question asked by the problem

Model Problem

Thelma and Laura start a lawn-mowing business and buy a lawnmower for $225. They plan to charge $15 to mow one lawn. What is the *minimum* number of lawns they need to mow if they wish to earn a profit of *at least* $750?

Solution:

(A) Let x represent the number of lawns they need to mow. Total profit is $15x - 225$. So,

(B) $15x - 225 \geq 750$

(C) $\underline{+225 +225}$

$$\frac{15x}{15} \geq \frac{975}{15}$$

$$x \geq 65$$

(D) They need to mow a minimum of 65 lawns.

Explanation of Steps:
(A) Translate each quantity into an expression in terms of a variable *[profit would be $15 per mowed lawn, less the $225 cost of the lawnmower, or 15x – 225]*.
(B) Determine from the problem how the expressions form an inequality *[the profit must be at least – greater than or equal to – $750]*.
(C) Write and solve the inequality.
(D) Be sure to answer the question asked by the problem.

Practice Problems

1. Find the smallest integer such that five less than twice the integer is greater than 23.	2. The larger of two integers is 7 times the smaller. The sum of the integers is at most 60. What are the largest two integers that can make these statements true?
3. Andy earns $5.95 per hour working after school. He needs at least $215 for his holiday shopping. How many hours must he work to reach his goal?	4. The cost per month of making n number of wooden toys is $3n + 30$. The income from selling n toys is $6n$. How many toys must the company make to make a profit (the income is greater than the cost)?
5. Parking charges at Superior Parking Garage are $5.00 for the first hour and $1.50 for each additional 30 minutes. If Margo has $12.50, what is the maximum amount of time she will be able to park her car at the garage?	6. Members of the band boosters are planning to sell programs at football games. The cost to print the programs is $150 plus $0.50 per program. They plan to sell each program for $2. How many programs must they sell to make a profit of at least $500?

REGENTS QUESTIONS

Multiple Choice

1. Tamara has a cell phone plan that charges $0.07 per minute plus a monthly fee of $19.00. She budgets $29.50 per month for total cell phone expenses without taxes. What is the maximum number of minutes Tamara could use her phone each month in order to stay within her budget?
   - (1) 150
   - (2) 271
   - (3) 421
   - (4) 692

2. An online music club has a one-time registration fee of $13.95 and charges $0.49 to buy each song. If Emma has $50.00 to join the club and buy songs, what is the maximum number of songs she can buy?
   - (1) 73
   - (2) 74
   - (3) 130
   - (4) 131

3. If five times a number is less than 55, what is the greatest possible integer value of the number?
   - (1) 12
   - (2) 11
   - (3) 10
   - (4) 9

4. Jason's part-time job pays him $155 a week. If he has already saved $375, what is the minimum number of weeks he needs to work in order to have enough money to buy a dirt bike for $900?
   - (1) 8
   - (2) 9
   - (3) 3
   - (4) 4

Constructed Response

5. A prom ticket at Smith High School is $120. Tom is going to save money for the ticket by walking his neighbor's dog for $15 per week. If Tom already has saved $22, what is the minimum number of weeks Tom must walk the dog to earn enough to pay for the prom ticket?

6. Peter begins his kindergarten year able to spell 10 words. He is going to learn to spell 2 new words every day. Write an inequality that can be used to determine how many days, *d*, it takes Peter to be able to spell *at least* 75 words. Use this inequality to determine the minimum number of whole days it will take for him to be able to spell *at least* 75 words.

7. Chelsea has $45 to spend at the fair. She spends $20 on admission and $15 on snacks. She wants to play a game that costs $0.65 per game. Write an inequality to find the maximum number of times, *x*, Chelsea can play the game. Using this inequality, determine the maximum number of times she can play the game.

IV.   LINEAR GRAPHS

Determining Whether a Point is on a Line

Key Terms and Concepts

A **line** consists of a set of **points**, each of which can be represented by an **ordered pair** stating its x-value and y-value as (x,y). This set of points represents the **solution set** of a **linear equation** involving the two variables, x and y.

One way to graph a line is to by **creating a table**. The first column will contain some sample x-values; I usually prefer to choose –2, –1, 0, 1, and 2. The second column is used to substitute the x-value into the equation in order to solve for y. The resulting y-values are written in the third column. The last column gives the corresponding ordered pairs of x- and y-values.

For example: We can graph $y = 3x + 1$ by using a table and drawing a line through the ordered pairs on a coordinate graph.

x-value	$y = 3x + 1$	y-value	(x,y)
–2	$y = 3(-2) + 1$	–5	(–2,–5)
–1	$y = 3(-1) + 1$	–2	(–1,–2)
0	$y = 3(0) + 1$	1	(0,1)
1	$y = 3(1) + 1$	4	(1,4)
2	$y = 3(2) + 1$	7	(2,7)

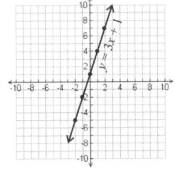

You can **determine whether a point is on a line** by substituting the x-value and y-value for the variables x and y in the equation and then checking if these values make the equation true.

For example: $(4,13)$ is on the line $y = 3x + 1$ because substituting 4 for x and 13 for y, we get $13 = 3(4) + 1$, which is true. This can also be done using the calculator:

4 [STO▸] [ALPHA] [X] [ENTER]   13 [STO▸] [ALPHA] [Y] [ENTER]

[ALPHA] [Y] [2nd] [TEST] [=] 3 [ALPHA] [X] + 1 [ENTER]

Model Problem
Does the line whose equation is $2y + 6 = 4x$ contain the point $(1,-1)$?

Solution:
   (A) $2y + 6 = 4x$ for $x = 1$, $y = -1$
   (B) $2(-1) + 6 = 4(1)$?
   (C) $4 = 4$, so Yes, $(1,-1)$ is in the solution set.

Explanation of Steps:
   (A) The first value in the ordered pair represents the value of x, the second is the value of y.
   (B) Substitute for x and y.
   (C) Evaluate both sides of the equation to determine if the equation is true.

Practice Problems

1. Does the point (3,7) lie on the line whose equation is $y = 3x - 2$?	2. Does the line whose equation is $y = \dfrac{1}{2}x + 5$ contain the point (4,9)?
3. Does the line whose equation is $y = 4x$ pass through the origin, (0,0)?	4. Does the point (–2,–4) lie on the line whose equation is $2y - 3x = -2$?
5. Determine if the ordered pair (–4, 3) is a solution of $4x - y = -13$.	6. Determine if the ordered pair (–2, –4) is a solution of $5x - 2y = -2$.
7. Determine if the ordered pair (–5, –1) is a solution of $2x - y = -11$.	8. Determine if the ordered pair (3, –2) is a solution of $4x = 3y + 18$.
9. The graph of the equation $2x + 6y = 4$ passes through point $(x,-2)$. What is the value of x?	10. If $(k,3)$ is a point on the line whose equation is $4x + y = -9$, what is the value of k?

REGENTS QUESTIONS

Multiple Choice

1. The graph of which equation passes through the point (0,0)?
    (1) $x = 3$ (3) $y = 3x + 3$
    (2) $y = 3$ (4) $y = 3x$

2. Point (k,–3) lies on the line whose equation is $x - 2y = -2$. What is the value of *k*?
    (1) –8   (3) 6
    (2) –6   (4) 8

3. Which point is on the line $4y - 2x = 0$?
    (1) (–2,–1)   (3) (–1,–2)
    (2) (–2,1)   (4) (1,2)

4. Which linear equation represents a line containing the point (1,3)?
    (1) $x + 2y = 5$ (3) $2x + y = 5$
    (2) $x - 2y = 5$ (4) $2x - y = 5$

5. Which point lies on the line whose equation is $2x - 3y = 9$?
    (1) (–1,–3)   (3) (0,3)
    (2) (–1,3)   (4) (0,–3)

6. Which point lies on the graph represented by the equation $3y + 2x = 8$?
    (1) (–2,7)   (3) (2,4)
    (2) (0,4)   (4) (7,–2)

7. Which set of coordinates is a solution of the equation $2x - y = 11$?
    (1) (–6,1)   (3) (0,11)
    (2) (–1,9)   (4) (2,–7)

8. If the point (5,k) lies on the line represented by the equation $2x + y = 9$, the value of *k* is
    (1) 1   (3) –1
    (2) 2   (4) –2

Constructed Response

9. A point on the graph of $y = 2x - 4$ has an *x*-coordinate of 3. Find the *y*-coordinate of this point.

Lines Parallel to Axes

Key Terms and Concepts

If a linear equation has only one variable (that is, x or y is equal to a constant), then it represents a line that is parallel to one of the axes.

If **y is equal to a constant**, the line is parallel to the x-axis and crosses the y-axis at that constant.
For example: $y = 3$ represents a line parallel to the x-axis but 3 units above it.

No matter what values we choose for x, the y value is always 3, as shown below.

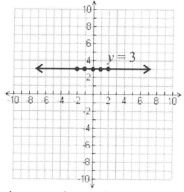

x-value	$y=3$	y-value	(x,y)
-2	$y=3$	3	$(-2,3)$
-1	$y=3$	3	$(-1,3)$
0	$y=3$	3	$(0,3)$
1	$y=3$	3	$(1,3)$
2	$y=3$	3	$(2,3)$

If **x is equal to a constant**, the line is parallel to the y-axis and crosses the x-axis at that constant.
For example: $x = -5$ represents a line parallel to the y-axis but 5 units to the left of it.
The line contains the points $(-5, y)$ where y is any real number.

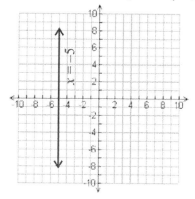

Model Problem
Write the equation of a line parallel to the x-axis but 10 units below it.

Solution:
$y = -10$

Explanation of Steps:
If a line is parallel to an axis, then the other variable is equal to a constant. *[A line parallel to the x-axis has y = constant; since it is 10 units below the axis, $y = -10$.]*

Practice Problems

1. Write the equation of the line parallel to the *y*-axis but 9 units to the right of it.	2. Write the equation of the line parallel to the *x*-axis but 1 unit above it.
3. Write the equation of the line that lies on the *y*-axis.	4. Write the equation of the line that lies on the *x*-axis.
5. At what point does the line whose equation is $x = 5$ intersect the *x*-axis, written as an ordered pair?	6. On a separate coordinate graph, graph the line whose equation is $x = 7$.
7. On a separate coordinate graph, graph the line whose equation is $y = -4$.	8. On a separate coordinate graph, graph the line whose equation is $y - 4 = -1$.

REGENTS QUESTIONS

Multiple Choice

1. Which equation represents a line parallel to the *x*-axis?

    (1) $x = 5$ (3) $x = \dfrac{1}{3}y$

    (2) $y = 10$ (4) $y = 5x + 17$

2. Which equation represents a line parallel to the *x*-axis?

    (1) $y = -5$ (3) $x = 3$

    (2) $y = -5x$ (4) $x = 3y$

3. Which equation represents a line parallel to the *y*-axis?

    (1) $x = y$ (3) $y = 4$

    (2) $x = 4$ (4) $y = x + 4$

4. Which equation represents a line parallel to the *y*-axis?

    (1) $y = x$ (3) $x = -y$

    (2) $y = 3$ (4) $x = -4$

5. Which equation represents the line that passes through the point (3,4) and is parallel to the *x*-axis?

    (1) $x = 4$ (3) $y = 4$

    (2) $x = -3$ (4) $y = -3$

6. Which equation represents a line that is parallel to the *y*-axis and passes through the point (4,3)?

    (1) $x = 3$ (3) $y = 3$

    (2) $x = 4$ (4) $y = 4$

7. Which equation represents a line that is parallel to the *y*-axis?

    (1) $x = 5$ (3) $y = 5$

    (2) $x = 5y$ (4) $y = 5x$

Finding Slope Given Two Points

Key Terms and Concepts

If you can imagine a person walking along a line *from left to right*, the **slope of the line** represents how steep the road is.

A **positive slope** would represent walking uphill and a **negative slope** would represent walking downhill. A **slope of zero** would mean the road is horizontal (parallel to the *x*-axis); it is neither uphill nor downhill. The person cannot walk from left to right along a vertical line (parallel to the *y*-axis), so we say that a vertical line has **no slope**. The larger the absolute value of the slope, the steeper the road: a slope of 3 is a steeper uphill climb than a slope of 1/3, and a slope of –3 is a steeper downhill descent than a slope of –1/3.

The slope is usually represented by the letter ***m***. Given two points on a line, we can determine, either graphically or algebraically, the slope of the line.

Finding the slope graphically:

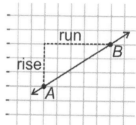

If we move from the left point to the right point, the slope $m = \dfrac{rise}{run}$.

The **rise** is how many units we need to travel up (positive) or down (negative), and the **run** is how many units we need to travel to the right.

For example: From point A to point B on the graph at right, the rise is 3 and the run is 5, so the slope is $\dfrac{3}{5}$.

Finding the slope algebraically:

We can think of the rise, or the number of y-value units we need to travel, as the **difference in the y-values**, and the run, or the number of x-value units we need to travel, as the **difference in the x-values**. If we name the coordinates of the two points (x_1, y_1) and (x_2, y_2), and we use the delta symbol Δ for the "change" or difference in values, the slope formula can be written as:

$$m = \frac{\Delta y}{\Delta x} = \frac{y_2 - y_1}{x_2 - x_1} \qquad \textit{[This formula is included on the reference sheet.]}$$

For example: The slope of the line through (1,3) and (2,6) is $m = \dfrac{y_2 - y_1}{x_2 - x_1} = \dfrac{6-3}{2-1} = \dfrac{3}{1} = 3$

If the two points lie on a horizontal line (parallel to the *x*-axis), the slope is zero.

For example: (2,5) and (3,5) lie on the horizontal line $y = 5$. $m = \dfrac{y_2 - y_1}{x_2 - x_1} = \dfrac{5-5}{3-2} = \dfrac{0}{1} = 0$

If the two points lie on a vertical line (parallel to the *y*-axis), there is no slope.

For example: (4,1) and (4,3) lie on the vertical line $x = 4$. To use the slope formula would result in a *denominator of zero*, which means that the slope is *undefined*.

Model Problem 1: *graphically*

What is the slope of the line passing through the points (2,4) and (6,6)?

Solution:

$$m = \frac{rise}{run} = \frac{2}{4} = \frac{1}{2}$$

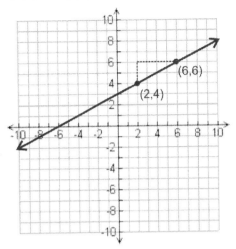

Explanation of Steps:

(A) Using the graph, first trace how many units you need to travel up (+) or down (−), and call this the rise *[from y=4 to y=6 is 2 units]*.

(B) Next, trace how many units you need to travel to the right, and call this the run *[from x=2 to x=6 is 4 units]*.

(C) Write $\frac{rise}{run}$ as a fraction, and reduce.

Practice Problems

1. What is the slope of line ℓ in the accompanying diagram? 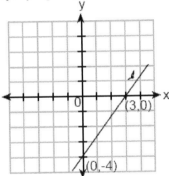	2. What is the slope of line ℓ in the accompanying diagram? 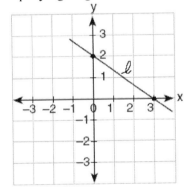
3. What is the slope of the line shown below? 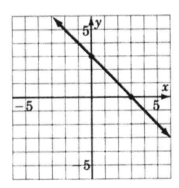	4. On a separate coordinate graph, draw a line through the points (−4,−3) and (1,2) and find the slope of the line.

89

Model Problem 2: *algebraically*
What is the slope of the line passing through the points (–3,4) and (1,2)?

Solution:

(A) $(x_1, y_1) = (-3, 4)$ $(x_2, y_2) = (1, 2)$

(B)   (C)   (D)   (E)

$$m = \frac{y_2 - y_1}{x_2 - x_1} = \frac{2 - 4}{1 - (-3)} = \frac{-2}{4} = -\frac{1}{2}$$

Explanation of Steps:
   (A) Label the coordinates of the two points. Each point has an x and y value, but the first point has subscripts of 1 and the second point has subscripts of 2.
   (B) Write the slope formula.
   (C) Substitute the coordinates as labeled.
   (D) Evaluate the numerator and denominator.
   (E) Simplify, if possible.

Practice Problems

5. What is the slope of the line that passes through points (1,3) and (5,13)?	6. What is the slope of the line that passes through points (3,–6) and (1,8)?
7. What is the slope of the line that passes through points (4,5) and (0,–3)?	8. What is the slope of the line that passes through (–4,–2) and (2,–2)?

REGENTS QUESTIONS

Multiple Choice

1. What is the slope of the line containing the points (3,4) and (-6,10)?

   (1) $\dfrac{1}{2}$ (3) $-\dfrac{2}{3}$

   (2) 2   (4) $-\dfrac{3}{2}$

2. What is the slope of the line that passes through the points (-6,1) and (4,-4)?

   (1) -2   (3) $-\dfrac{1}{2}$

   (2) 2   (4) $\dfrac{1}{2}$

3. What is the slope of the line that passes through the points (2,5) and (7,3)?

   (1) $-\dfrac{5}{2}$ (3) $\dfrac{8}{9}$

   (2) $-\dfrac{2}{5}$ (4) $\dfrac{9}{8}$

4. What is the slope of the line that passes through the points (-5,4) and (15,-4)?

   (1) $-\dfrac{2}{5}$ (3) $-\dfrac{5}{2}$

   (2) 0   (4) undefined

5. In the diagram below, what is the slope of the line passing through points *A* and *B*?

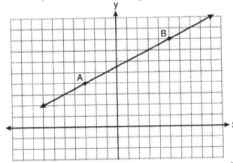

   (1) –2   (3) $-\dfrac{1}{2}$

   (2) 2   (4) $\dfrac{1}{2}$

6. What is the slope of the line that passes through the points (3,5) and (–2,2)?

   (1) $\dfrac{1}{5}$ (3) $\dfrac{5}{3}$

   (2) $\dfrac{3}{5}$ (4) 5

91

7. What is the slope of the line passing through the points A and B, as shown on the graph below?

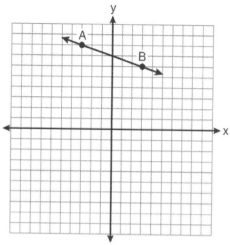

(1) –3   (3) 3

(2) $-\dfrac{1}{3}$ (4) $\dfrac{1}{3}$

8. What is the slope of the line passing through the points (–2,4) and (3,6)?

(1) $-\dfrac{5}{2}$ (3) $\dfrac{2}{5}$

(2) $-\dfrac{2}{5}$ (4) $\dfrac{5}{2}$

9. What is the slope of the line that passes through the points (2,–3) and (5,1)?

(1) $-\dfrac{2}{3}$ (3) $-\dfrac{4}{3}$

(2) $\dfrac{2}{3}$ (4) $\dfrac{4}{3}$

10. What is the slope of the line that passes through the points (4,–7) and (9,1)?

(1) $\dfrac{5}{8}$ (3) $-\dfrac{6}{13}$

(2) $\dfrac{8}{5}$ (4) $-\dfrac{13}{6}$

11. What is the slope of a line that passes through the points (–2,–7) and (–6,–2)?

(1) $-\dfrac{4}{5}$ (3) $\dfrac{8}{9}$

(2) $-\dfrac{5}{4}$ (4) $\dfrac{9}{8}$

Finding Slope Given an Equation

Key Terms and Concepts

The equation of a line is most commonly written in **slope-intercept form**: $y = \mathbf{m}x + \mathbf{b}$.
In this form, **m** (the coefficient of x) represents the **slope** of the line and **b** is the **y-intercept**. The y-intercept is the value of y at the point where the line intersects the y-axis.

Transforming an equation into slope-intercept form
If the equation is not already in slope-intercept form, you will need to transform the equation by solving for y in terms of x.
For examples: To transform the equations $y + 2x = 4$ and $3y = 2x - 9$,

$$y + 2x = 4$$
$$\underline{-2x \quad -2x}$$
$$y = -2x + 4$$

$$\frac{3y}{3} = \frac{2x - 9}{3}$$
$$y = \frac{2}{3}x - 3$$

Model Problem
What is the slope of the line whose equation is $2y - 3x = x + 2$?

Solution:

$$2y - 3x = x + 2$$
(A) $$\underline{+3x \quad +3x}$$
$$\frac{2y}{2} = \frac{4x + 2}{2}$$
$$y = 2x + 1$$
(B)   Slope is 2

Explanation of Steps:

(A) If the equation is not already in slope-intercept form, transform it by solving for y in terms of x.
   [Add $3x$ to both sides, then divide each by 2.]

(B) For an equation in slope-intercept form, the slope is the coefficient of x *[the slope is 2]*.

Practice Problems

1. What is the slope of a line whose equation is $y = \dfrac{2}{5}x - 5$?	2. What is the slope of the line whose equation is $y - 3x = 1$?
3. What is the slope of the line whose equation is $2y = 5x + 4$?	4. What is the slope of the linear equation $5y - 10x = -15$?
5. What is the slope of the line whose equation is $3x - 2y = 12$?	6. What is the slope of the line whose equation is $3x - 4y - 16 = 0$?

REGENTS QUESTIONS

Multiple Choice

1. What is the slope of the line whose equation is $3x - 7y = 9$?

    (1) $-\dfrac{3}{7}$ (3) $-\dfrac{7}{3}$

    (2) $\dfrac{3}{7}$ (4) $\dfrac{7}{3}$

2. The line represented by the equation $2y - 3x = 4$ has a slope of

    (1) $-\dfrac{3}{2}$ (3) 3

    (2) 2   (4) $\dfrac{3}{2}$

3. What is the slope of the line represented by the equation $4x + 3y = 12$?

    (1) $\dfrac{4}{3}$ (3) $-\dfrac{3}{4}$

    (2) $\dfrac{3}{4}$ (4) $-\dfrac{4}{3}$

4. What is the slope of a line represented by the equation $2y = x - 4$?

    (1) 1   (3) -1

    (2) $\frac{1}{2}$ (4) $-\frac{1}{2}$

Graphing a Linear Equation

Key Terms and Concepts

Graphing an equation in slope-intercept form
If given an equation in the form $y = mx + b$, the line can be graphed by following these steps:

  (A) Use the y-intercept to plot the point (0,b) on the y-axis.
  (B) Use the slope to determine at least two more points.
  (C) Draw a line through the points and label the line with the equation.
For example:

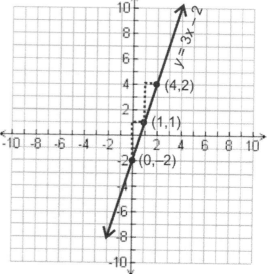

For the equation $y = 3x - 2$,

the slope $m = 3$ and the y-intercept $b = -2$.
The y-intercept of –2 gives us the starting point, (0,–2).

Since the slope $m = 3 = \dfrac{3}{1} = \dfrac{rise}{run}$,

use the rise of 3 and run of 1 to get two more points,
(1,1) and (2,4).

Some students like to remember that **b** tells us a point
to *begin* graphing the line and the **m** tells us how to
move to find other points on the line.

We can also **graph the line using the calculator**:
  1. Write the equation in slope-intercept form.
  2. Enter the equation by pressing the [Y=] button,
      then the right side of the equation and [ENTER].
      Use [ALPHA] [X] to enter the variable x.
  3. Press [ZOOM] [ZStandard] [ENTER].
For example: For the equation $y = 3x - 2$, enter

  [Y=] 3 [ALPHA] [X] – 2 [ENTER] [ZOOM] [ZStandard] [ENTER].

Note: The **ZStandard** function on the calculator sets the grid size to 20 by 20 units centered at
the origin. For graphs that may not display well in the standard grid size, you may need to adjust
the **Window** size. Press [WINDOW] and enter values for [Xmin] and [Xmax], and also values
for [Ymin] and [Ymax]. Then press [GRAPH] instead of [ZOOM] [ZStandard].

To view a table of points on the line: Press [2nd] [TABLE]. You can then scroll up or down with
the arrow keys [▲][▼] to see more points.

96

REGENTS QUESTIONS

Multiple Choice

1. Which equation is represented by the graph below?

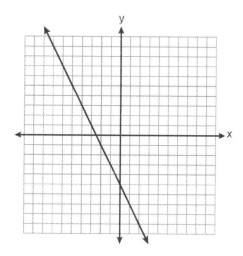

    (1) $2y + x = 10$ (3) $-2y = 10x - 4$

    (2) $y - 2x = -5$ (4) $2y = -4x - 10$

Constructed Response

2. On the set of axes below, draw the graph of the equation $y = -\dfrac{3}{4}x + 3$.

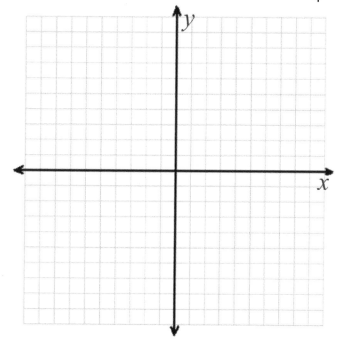

    Is the point (3,2) a solution to the equation? Explain your answer based on the graph drawn.

Equations of Parallel Lines

Key Terms and Concepts

If two distinct lines have the **same slope**, they are **parallel**. So, to determine whether two lines are parallel, write each equation in *slope-intercept form* to determine whether the slopes (the coefficients of *x*) are the same for both equations.

Model Problem

The equations of two distinct lines are $y = 3x - 6$ and $2y = 3x + 6$. Are the lines parallel?

Solution:

(A) For $y = 3x - 6$, the slope $m = 3$.

Solving $2y = 3x + 6$ for *y*:

$$\frac{2y}{2} = \frac{3x + 6}{2}$$

$y = \dfrac{3}{2}x + 3$, so the slope $m = \dfrac{3}{2}$.

(B) The lines are *not* parallel because the slopes are not equal.

Explanation of Steps:

(A) Write each equation in slope-intercept form to determine the slope of each line.

[The first equation is already in slope-intercept form, $y = mx + b$, so the slope $m = 3$.

The second equation needed to be transformed, resulting in a slope of $\frac{3}{2}$.]

(B) If the slopes are equal, the lines are parallel.

[These slopes are 3 and $\frac{3}{2}$, so they are not parallel.]

Practice Problems

1. Line ℓ has an equation of $y = -2x - 5$. Write the equation of a line that is parallel to line ℓ but has a y-intercept of 2.	2. Line ℓ has an equation of $y = \dfrac{1}{2}x + 2$. Write the equation of a line that is parallel to line ℓ but passes through the origin.
3. Which equation below represents a line that is parallel to the line, $y = -x + 4$? $2y + 2x = 6$ $2y - x = 6$	4. Which equation below represents a line that is parallel to the line, $4x + 6y = 5$? $-3y = 2x + 5$ $-6y + 4x = 5$

REGENTS QUESTIONS

Multiple Choice

1. An equation of the line whose y-intercept is -4 and whose graph is parallel to the graph of $y = -2x + 7$ is

    (1) $y = \dfrac{1}{2}x - 4$

    (2) $y = -2x - 4$

    (3) $y = \dfrac{1}{4}x + 7$

    (4) $y = -4x + 7$

2. Which equation represents a line that is parallel to the line $y = -4x + 5$?

    (1) $y = -4x + 3$

    (2) $y = -\dfrac{1}{4}x + 5$

    (3) $y = \dfrac{1}{4}x + 3$

    (4) $y = 4x + 5$

3. Which equation represents a line that is parallel to the line $y = 3 - 2x$?

    (1) $4x + 2y = 5$

    (2) $2x + 4y = 1$

    (3) $y = 3 - 4x$

    (4) $y = 4x - 2$

4. Which equation represents a line parallel to the graph of $2x - 4y = 16$?

    (1) $y = \dfrac{1}{2}x - 5$

    (2) $y = -\dfrac{1}{2}x + 4$

    (3) $y = -2x + 6$

    (4) $y = 2x + 8$

5. The graphs of the equations $y = 2x - 7$ and $y - kx = 7$ are parallel when k equals

    (1) -2

    (2) 2

    (3) -7

    (4) 7

6. Which equation represents a line that is parallel to the line whose equation is $2x - 3y = 9$?

    (1) $y = \dfrac{2}{3}x - 4$

    (2) $y = -\dfrac{2}{3}x + 4$

    (3) $y = \dfrac{3}{2}x - 4$

    (4) $y = -\dfrac{3}{2}x + 4$

7. Which equation represents a line that is parallel to the line whose equation is $y = -3x - 7$?

    (1) $y = -3x + 4$

    (2) $y = -\dfrac{1}{3}x - 7$

    (3) $y = \dfrac{1}{3}x + 5$

    (4) $y = 3x - 2$

Writing a Linear Equation Given a Point and Slope

Key Terms and Concepts

If given the **slope** of a line and the coordinates of **one point**, substitute the slope for m and the coordinates of the point for x and y in the general **slope-intercept form**, $y = mx + b$. This will allow us to solve for b. Once m and b are known, the equation of the line may be written.

An alternative method uses the general **point-slope form** of a linear equation:
$$y - y_1 = m(x - x_1)$$ where m is the slope and (x_1, y_1) is a point on the line.

Using this form, we can substitute the given slope for m and the coordinates of the point for x_1 and y_1. If we want to transform the resulting equation into slope-intercept form, we simply need to solve for y in terms of x.

Model Problem

Write the equation of the line through $(1, -3)$ with a slope of 2.

Solution:

(A) $y = mx + b$

(B) $-3 = 2(1) + b$

(C) $-3 = 2 + b$

$\quad -5 = b$

(D) $y = 2x - 5$

Explanation of Steps:

(A) Write the slope-intercept form of an equation.

(B) Substitute the given values *[x = 1, y = −3, m = 2]*.

(C) Solve for b.

(D) Write the equation using the given slope m and the value of b.

Practice Problems

1. Write an equation of a line through point $(1,4)$ with a slope of 2.	2. Write an equation of the line passing through point $(-6, 5)$ with a slope of 5.
3. Write an equation of a line that passes through $(-3,2)$ and has a slope of $\frac{1}{3}$.	4. Write an equation of the line passing through point $(8, -3)$ with a slope of $\frac{3}{4}$.

REGENTS QUESTIONS

Multiple Choice

1. What is an equation of the line that passes through the point $(4,-6)$ and has a slope of -3?

      (1) $y=-3x+6$ (3) $y=-3x+10$

      (2) $y=-3x-6$ (4) $y=-3x+14$

2. What is an equation of the line that passes through the point $(3,-1)$ and has a slope of 2?

      (1) $y=2x+5$ (3) $y=2x-4$

      (2) $y=2x-1$ (4) $y=2x-7$

3. Which equation represents the line that passes through the point $(1,5)$ and has a slope of -2?

      (1) $y=-2x+7$ (3) $y=2x-9$

      (2) $y=-2x+11$ (4) $y=2x+3$

4. Which equation represents a line that has a slope of $\frac{3}{4}$ and passes through the point $(2,1)$?

      (1) $3y=4x-5$ (3) $4y=3x-2$

      (2) $3y=4x+2$ (4) $4y=3x+5$

5. What is an equation of the line that passes through the point $(-2,-8)$ and has a slope of 3?

      (1) $y=3x-2$ (3) $y=3x+2$

      (2) $y=3x-22$ (4) $y=3x+22$

6. What is the equation of the line that passes through the point $(3,-7)$ and has a slope of $-\frac{4}{3}$?

      (1) $y=-\frac{4}{3}x+3$ (3) $y=\frac{37}{3}x-\frac{4}{3}$

      (2) $y=-\frac{4}{3}x-3$ (4) $y=-\frac{59}{9}x-\frac{4}{3}$

Constructed Response

7. A line having a slope of $\frac{3}{4}$ passes through the point $(-8,4)$. Write the equation of this line in slope-intercept form.

Writing a Linear Equation Given Two Points

Key Terms and Concepts

Given two points on a line, we can write the equation of the line in **slope-intercept form**. First, find the slope m. Then, substitute the slope and one of the point's coordinates into the general equation, $y = mx + b$, in order to solve for b. Once you have m and b, you can write the equation in slope-intercept form.

An alternative method uses the point-slope form of a linear equation. The general form of the point-slope equation is $y - y_1 = m(x - x_1)$. Find the slope m. Then substitute one of the point's coordinates for x_1 and y_1.

Model Problem
Write an equation of the line that passes through (3,–2) and (6,4).

Solution:

(A) $m = \dfrac{y_2 - y_1}{x_2 - x_1} = \dfrac{4 - (-2)}{6 - 3} = \dfrac{6}{3} = 2$

(B) $y = mx + b$

(C) $4 = 2(6) + b$

(D) $4 = 12 + b$

$-8 = b$

(E) $y = 2x - 8$

Explanation of Steps:

(A) find the slope of the line
(B) write the general slope-intercept form of an equation
(C) substitute one point's coordinates *[(6,4)]* for x and y, and the slope *[2]* for m
(D) solve for b
(E) write the resulting equation using the calculated values of m and b

Practice Problems

1. Write an equation of the line that passes through the points (1,2) and (5,6).	2. Write an equation of the line that passes through the points (2,–1) and (3,4).
3. Write an equation of the line that passes through the points (–3,0) and (3,–2).	4. Write an equation of the line that passes through the points (–2,4) and (2,4).

REGENTS QUESTIONS

Multiple Choice

1. What is an equation for the line that passes through the coordinates $(2,0)$ and $(0,3)$?

   (1) $y = -\dfrac{3}{2}x + 3$ (3) $y = -\dfrac{2}{3}x + 2$

   (2) $y = -\dfrac{3}{2}x - 3$ (4) $y = -\dfrac{2}{3}x - 2$

2. What is an equation of the line that passes through the points $(3,-3)$ and $(-3,-3)$?

   (1) $y = 3$ (3) $y = -3$
   (2) $x = -3$ (4) $x = y$

3. Which equation represents the line that passes through the points $(-3,7)$ and $(3,3)$?

   (1) $y = \dfrac{2}{3}x + 1$ (3) $y = -\dfrac{2}{3}x + 5$

   (2) $y = \dfrac{2}{3}x + 9$ (4) $y = -\dfrac{2}{3}x + 9$

4. What is an equation of the line that passes through the points $(1,3)$ and $(8,5)$?

   (1) $y + 1 = \dfrac{2}{7}(x + 3)$ (3) $y - 1 = \dfrac{2}{7}(x + 3)$

   (2) $y - 5 = \dfrac{2}{7}(x - 8)$ (4) $y + 5 = \dfrac{2}{7}(x - 8)$

Constructed Response

5. Write an equation that represents the line that passes through the points $(5, 4)$ and $(-5, 0)$.

Graphing Inequalities

Key Terms and Concepts

To graph a linear inequality, start by isolating (solving for) the variable y. Then, consider the graph of the equation that would result if the inequality symbol was replaced by an equal sign.

The points on this line are *included in the solution set* if the inequality symbol is $\leq$ or $\geq$, but *not included in the solution set* if the inequality symbol is $<$ or $>$. To show inclusion ($\leq$ or $\geq$), draw a **solid line**; otherwise, draw a **dashed line**.

The line divides the plane into two parts. The part above the line includes all points where $y > mx + b$ and the part below the line includes all points where $y < mx + b$. So, if the inequality starts with $y >$ or $y \geq$, shade **above the line**; if it starts with $y <$ or $y \leq$, then shade **below the line**. Points in the shaded area are *included in the solution set*.

Special case: vertical lines

If the only variable in the inequality is x, solve for x and graph the corresponding vertical line (either solid or dashed according to the same rules). Then shade to the **right of the line** for $x >$ or $x \geq$, or to the **left of the line** for $x <$ or $x \leq$.

For example: The graph of $x \geq 5$ will have a solid vertical line at $x = 5$ and shading to the right of the line.

 Inequalities can also be graphed on the calculator. Follow the same steps as entering an equation, but then change the symbol to the left of $\boxed{Y_1 =}$ from a line $\boxed{\diagdown}$ to a "shade above" $\boxed{\blacksquare}$ or "shade below" $\boxed{\blacksquare}$ symbol by moving to the line symbol and pressing $\boxed{\text{ENTER}}$ repeatedly. You will still need to know whether to use a solid or dashed line when drawing your graph on paper, depending on the inequality symbol.

Model Problem

Graph the inequality $-4y < 5x - 20$. Is the point (1,2) in the solution set?

Solution:

(A) Solving for y,

$$\frac{-4y}{-4} < \frac{5x - 20}{-4}$$

$$y > -\frac{5}{4}x + 5$$

(C) (1,2) is not in the solution set.

(B)

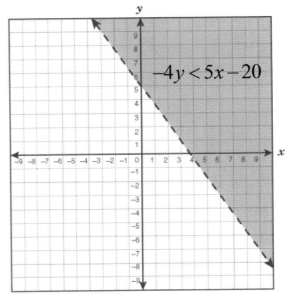

Explanation of Steps:

(A) Solve the inequality for y. *[Remember that multiplying or dividing both sides of an inequality by a negative value requires that you reverse the inequality symbol.]*

(B) Use the y-intercept and slope to graph the line. Use a solid line for $\leq$ or $\geq$, or a dashed line for $<$ or $>$. Shade above the line if y is $>$ or $\geq$, or below the line is y is $<$ or $\leq$. *[The $>$ means dashed and shaded above.]*

(C) If a point lies on a solid line or in a shaded area, it is in the solution set; otherwise, it is not. *[(1,2) lies in the unshaded area below the line, so it is not a solution.]*

 You could also graph $y > -\frac{5}{4}x + 5$ on the calculator:

1. Press $\boxed{Y=}$ then the left arrow $\boxed{\blacktriangleleft}$ twice until the cursor is over the $\boxed{\diagdown}$ symbol.

2. Since the inequality starts with "$y >$" press the $\boxed{ENTER}$ two times until the shade above $\boxed{\ }$ symbol appears. Move the cursor back to the right, after the $\boxed{=}$ sign.

3. Enter the right side of the inequality, $-5 \div 4 \boxed{ALPHA} [X] + 5 \boxed{ENTER}$.

4. Press $\boxed{ZOOM}$ $\boxed{ZStandard}$.

Practice Problems

1. Write an inequality that is represented by the graph below. 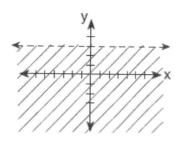	2. Write an inequality that is represented by the graph below. 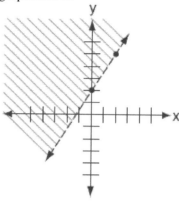
3. Graph the inequality $y \geq 4$ on a separate coordinate graph.	4. Graph the inequality $x < -1$ on a separate coordinate graph.
5. Graph the inequality $y > x - 2$ on a separate coordinate graph.	6. Graph the inequality $y \leq -\dfrac{2}{3}x + 5$ on a separate coordinate graph.
7. Graph the inequality $x + y \leq -3$ on a separate coordinate graph.	8. Graph the inequality $x - y \leq -1$ on a separate coordinate graph.
9. Graph the inequality $2y - 6x > 10$ on a separate coordinate graph.	10. Graph the inequality $9 - x \geq 3y$ on a separate coordinate graph.

REGENTS QUESTIONS

Multiple Choice

1. Which inequality is represented by the graph below?

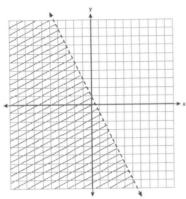

(1) $y < 2x + 1$ (3) $y < \dfrac{1}{2}x + 1$

(2) $y < -2x + 1$ (4) $y < -\dfrac{1}{2}x + 1$

2. Which graph represents the solution of $3y - 9 \le 6x$?

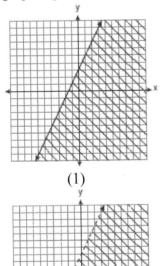

(1)

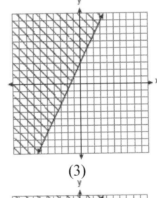

(3)

(2)

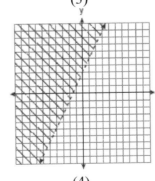

(4)

3. Which quadrant will be completely shaded in the graph of the inequality $y \leq 2x$?
   - (1) Quadrant I
   - (2) Quadrant II
   - (3) Quadrant III
   - (4) Quadrant IV

4. Which graph represents the inequality $y > 3$?

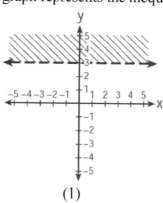

(1)

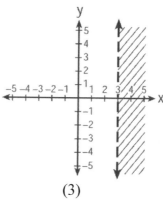

(3)

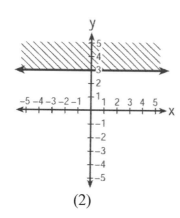

(2)

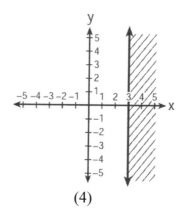

(4)

5. The diagram below shows the graph of which inequality?

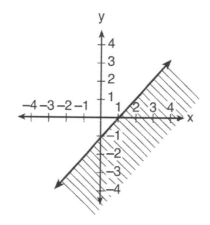

- (1) $y > x - 1$
- (2) $y \geq x - 1$
- (3) $y < x - 1$
- (4) $y \leq x - 1$

6. Which graph represents the inequality $y \geq x + 3$?

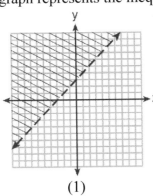

(1)

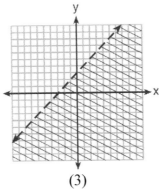

(3)

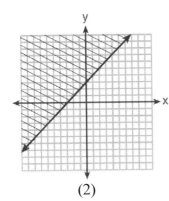

(2)

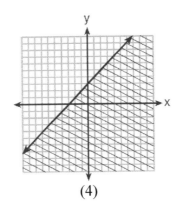

(4)

7. Which graph represents the solution of $2y + 6 > 4x$?

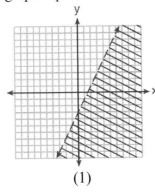

(1)

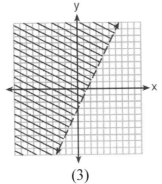

(3)

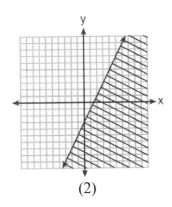

(2)

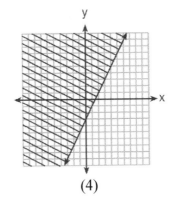

(4)

Constructed Response

8. Graph the solution set for the inequality $4x - 3y > 9$ on the set of axes below. Determine if the point $(1, -3)$ is in the solution set. Justify your answer.

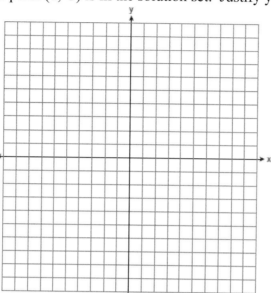

V.   LINEAR SYSTEMS

Solving Systems of Equations Algebraically

Key Terms and Concepts

A **linear system of equations** consists of two equations in two variables. A solution for a linear system is a set of one value for each variable that solves both equations at the same time. If the variables are x and y, the solution is an ordered pair which represents the point where the two equations' lines intersect on a coordinate plane.

There are two common methods used to solve linear systems algebraically: the addition method and the substitution method. In both methods, we aim to develop a new equation in which one of the two variables has been eliminated.

The **addition method (elimination method)** depends on the fact that if two equations are true, both the left sides and the right sides of each equation can be added to create a new true equation:

If $a = b$ and $c = d$, then $a + c = b + d$ or, written vertically:

$$\begin{aligned} a &= b \\ c &= d \\ \hline a+c &= b+d \end{aligned}$$

It also depends on the fact that we can multiply both sides of an equation by the same value and the equation remains true:

If $a = b$, then $ma = mb$

The goal in the addition method is to eliminate a variable by adding terms whose coefficients for that variable are additive inverses.

For example:
$$\begin{aligned} 2x - y &= 2 \\ x + y &= 4 \\ \hline 3x\quad &= 6 \end{aligned}$$
Adding $-y$ and $+y$ eliminates the variable y, allowing us to solve a simpler equation in one variable, $3x = 6$, or $x = 2$.

In the addition method, if two variable terms (in the same variable) are not already additive inverses, they can be made into additive inverses by multiplying one or both equations by values that will change the coefficients into inverses.

For example:
$$\begin{array}{ll} 5a + b = 13 & (\times 3) \\ 4a - 3b = 18 & \rightarrow \end{array} \qquad \begin{aligned} 15a + 3b &= 39 \\ 4a - 3b &= 18 \\ \hline 19a\quad &= 57 \end{aligned}$$

In the **substitution method**, an equation needs to have one of the variables expressed in terms of the other, or we will need to solve for one of the variables. Once we have an equation where a variable is equal to an expression, we can substitute that expression for that variable in the other equation.

For example:
$$\begin{aligned} y &= x + 1 \\ x + 2y &= 17 \quad \rightarrow \quad x + 2(x+1) = 17 \end{aligned}$$

In either method, once we know the value of one variable, we can substitute the known value into either original equation to solve for the other variable.

Model Problem 1: *addition method*

Solve the following system of equations using the addition method:

$$4a + 2b = 22$$

$$-4a + 3b = 3$$

Solution:

$$4a + 2b = 22$$
$$\underline{-4a + 3b = 3}$$

(A) $5b = 25$

(B) $b = 5$

(C) $4a + 2(5) = 22$

(D) $4a + 10 = 22$

$$4a = 12$$

$$a = 3$$

(E) Solution: $a = 3, \ b = 5$

Explanation of Steps:

(A) If the coefficients for one of the variables are additive inverses [$4a$ *and* $-4a$], add the equations to derive a new equation without that variable.

(B) Now that we have an equation with only one variable [*b*], solve for that one variable.

(C) Substitute this solution [*b=5*] into either one of the original equations.

(D) Solve for the other variable [a].

(E) Write the solution by stating the values of both variables.

Practice Problems

1. Solve the following system of equations for *x* and *y*: $$3x - y = 8$$ $$x + y = 4$$	2. Solve the following system of equations for *x* and *y*: $$2x - 3y = 19$$ $$3x + 3y = 21$$
3. Solve the following system of equations for *x* and *y*: $$2x - 4y = 12$$ $$-2x + y = -9$$	4. Solve the following system of equations for *x* and *y*: $$3x + y = 0$$ $$-x - y = -4$$

Model Problem 2: *addition method with multipliers*

Solve the following system of equations using the addition method:

$$5x + 8y = 1$$
$$3x + 4y = -1$$

Solution:

$$
\begin{array}{l}
\qquad\qquad\qquad\quad\text{(A)}\qquad\text{(B)} \\
5x + 8y = 1 \qquad\rightarrow\qquad 5x + 8y = 1 \\
3x + 4y = -1 \quad\times(-2)\quad \underline{-6x - 8y = 2} \\
\qquad\qquad\qquad\qquad\qquad -x \quad\;\; = 3 \\
\qquad\qquad\qquad\qquad\qquad\quad x = -3
\end{array}
$$

(C)
$$5(-3) + 8y = 1$$
$$-15 + 8y = 1$$
$$8y = 16$$
$$y = 2$$

(D) Solution: $(-3, 2)$

Explanation of Steps:

(A) If neither the x terms nor the y terms are additive inverses of each other, multiply one (or each) of the equations by a value to turn them into inverses *[to change the "y" terms into inverses, 8y and –8y, we can multiply the second equation by –2]*.

(B) Adding the equations eliminates the inverses and will now give us a new equation in one variable. Solve for that variable.

(C) Then, substitute the solution *[x = –3]* into either one of the original equations *[the first equation was used here]*, allowing you to solve for the other variable *[y = 2]*.

(D) The solution for variables x and y may be written as an ordered pair, (x,y).

Practice Problems

5. Solve the following system of equations for x and y: $$3x + 2y = 4$$ $$-2x + 2y = 24$$	6. What is the value of y in the following system of equations? $$2x + 3y = 6$$ $$2x + y = -2$$
7. Solve the following system of equations for x and y: $$-3x + 4y = 11$$ $$6x - 5y = -16$$	8. Solve the following system of equations for x and y: $$3x + 4y = 9$$ $$5x + 6y = 21$$

Model Problem 3: *substitution method*

Solve the following system of equations using the substitution method:

$$3x - y = 16$$
$$y = x - 8$$

Solution:

(A) $3x - (x - 8) = 16$

(B) $3x - x + 8 = 16$

$$2x + 8 = 16$$
$$2x = 8$$
$$x = 4$$

(C) $y = (4) - 8$

$$y = -4$$

(D) *Solution* : $(4, -4)$

Explanation of Steps:

(A) If one equation already has one of the variables isolated, as in *variable = expression*, substitute this expression for this variable in the other equation. *[$y = x - 8$ already has y expressed in terms of x, so substitute the expression $x - 8$ for the y in the first equation: $3x - y = 16$ becomes $3x - (x - 8) = 16$.]* It is always safest to use parentheses around the expression whenever you perform a substitution.

(B) Solve the equation for one variable.

(C) Substitute the solution found in step (B) into either original equation *[substitute 4 for x]*.

(D) Solve the equation for the other variable.

(E) State the solution.

Practice Problems

9. Solve the following system of equations for x and y:	10. Solve the following system of equations for x and y:
$$y = 4x - 10$$ $$y = 5 - x$$	$$x = y - 2$$ $$y = 10 - 3x$$
11. Solve the following system of equations for x and y:	12. Solve the following system of equations for x and y:
$$y = 9 - 2x$$ $$3y - 2x = 11$$	$$7x + 3y = 68$$ $$x - 4y = -8$$

REGENTS QUESTIONS

Multiple Choice

1. What is the value of y in the following system of equations?
$$2x + 3y = 6$$
$$2x + y = -2$$
   (1) 1   (3) –3
   (2) 2   (4) 4

2. What is the value of the y-coordinate of the solution to the system of equations $x + 2y = 9$ and $x - y = 3$?
   (1) 6   (3) 3
   (2) 2   (4) 5

3. What is the value of the y-coordinate of the solution to the system of equations $x - 2y = 1$ and $x + 4y = 7$?
   (1) 1   (3) 3
   (2) -1   (4) 4

4. What is the solution of the system of equations $c + 3d = 8$ and $c = 4d - 6$?
   (1) $c = -14, d = -2$ (3) $c = 2, d = 2$
   (2) $c = -2, d = 2$ (4) $c = 14, d = -2$

5. What is the value of the y-coordinate of the solution to the system of equations $2x + y = 8$ and $x - 3y = -3$?
   (1) –2   (3) 3
   (2) 2   (4) –3

6. What is the solution of the system of equations $2x - 5y = 11$ and $-2x + 3y = -9$?
   (1) (–3,–1)   (3) (3,–1)
   (2) (–1,3)   (4) (3,1)

7. Using the substitution method, Ken solves the following system of equations algebraically.
$$2x - y = 5$$
$$3x + 2y = -3$$
   Which equivalent equation could Ken use?
   (1) $3x + 2(2x - 5) = -3$ (3) $3\left(y + \dfrac{5}{2}\right) + 2y = -3$
   (2) $3x + 2(5 - 2x) = -3$ (4) $3\left(\dfrac{5}{2} - y\right) + 2y = -3$

8. What is the solution of the system of equations below?

$$2x + 3y = 7$$
$$x + y = 3$$

(1) $(1,2)$ (3) $(4,-1)$

(2) $(2,1)$ (4) $(4,1)$

9. What is the value of x in the solution of the system of equations $3x + 2y = 12$ and $5x - 2y = 4$?

(1) 8   (3) 3

(2) 2   (4) 4

10. Which system of equations has the same solution as the system below?

$$2x + 2y = 16$$
$$3x - y = 4$$

(1) $2x + 2y = 16$ (3) $x + y = 16$
$\quad\;\, 6x - 2y = 4$ $3x - y = 4$

(2) $2x + 2y = 16$ (4) $6x + 6y = 48$
$\quad\;\, 6x - 2y = 8$ $6x + 2y = 8$

Constructed Response

11. Solve the following system of equations for x:

$$3x + y = 7$$
$$2x - y = 8$$

12. Solve the following system of equations algebraically:

$$\frac{x}{y+1} = \frac{2}{3}$$
$$x + y = 9$$

13. Solve the following system of equations for x:

$$2x + y = 10$$
$$3x = y$$

14. Solve the following system of equations algebraically:

$$2x + 3y = -6$$
$$5x + 2y = 7$$

15. Solve the following system of equations algebraically:

$$y = 4x - 1$$
$$3x + 2y = 20$$

16. Solve the following system of equations algebraically:
$$4x = 5y + 35$$
$$6x + 7y = 9$$

17. Solve the following system of equations algebraically:
$$3x + 2y = 4$$
$$4x + 3y = 7$$

18. Solve the following system of equations algebraically for y:
$$2x + 2y = 9$$
$$2x - y = 3$$

Solving Systems of Equations Graphically

Key Terms and Concepts

To solve a system of linear equations **graphically**, simply graph the two equations as lines in the same coordinate plane. The point (if any) where the two lines **intersect** is the solution.

For example: The lines $y = -x + 5$ and $y = 2x - 4$ intersect at (3,2).

Therefore, the solution for the system of equations is $x = 3$ and $y = 2$.

If the two lines are parallel, they never intersect, and so there is no solution. If the two lines are identical (coincide), then there are infinitely many solutions.

 Using a calculator, you can solve a system of equations by graphing both equations and then using the intersect feature.

For example: To solve the system, $y = -x + 5$ and $y = 2x - 4$, graphically,

Press   [Y=] − [ALPHA] [X] + 5 [ENTER]
          2 [ALPHA] [X] − 4 [ENTER]
          [2nd] [CALC] |intersect|
          First curve? [ENTER] Second curve? [ENTER] Guess? [ENTER]

Model Problem

Solve the following system of equations graphically:

$$x - y = -1$$
$$y = -3x + 9$$

Solution:

(A) $x - y = -1$

$\quad\underline{-x \qquad -x}$

$\quad \dfrac{-y = -x - 1}{-1 \qquad -1}$

$\quad y = x + 1$

(B)

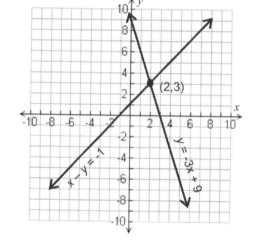

(2,3)

(C) Solution: (2, 3)

Explanation of Steps:

(A) If either equation is not already in slope-intercept form, transform the equation by solving for y in terms of x.

(B) Graph both equations on the same coordinate plane, labeling each line.

(C) The point of intersection is the solution to the system of equations.

[A solution of (2,3) means $x = 2$, $y = 3$ solves both equations simultaneously.]

Practice Problems

1. Solve the system of equations graphically: $y = 3x - 2$ $y = -x - 6$	2. Solve the system of equations graphically: $x + y = 2$ $x - y = 4$

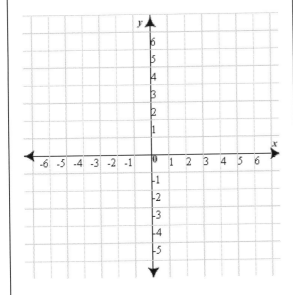

3. Solve the system of equations graphically: $3x - 5y = 15$ $y = 2x + 4$	4. Solve the system of equations graphically: $y = \dfrac{2}{3}x + 5$ $x + 3y = -3$

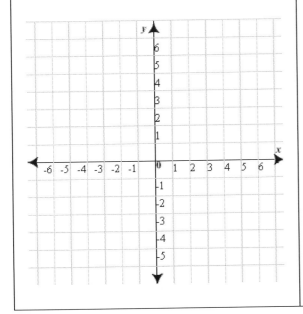

REGENTS QUESTIONS

Multiple Choice

1. If the lines whose equations are $x = -2$ and $y = 3$ were graphed on the same set of coordinate axes, their point of intersection would be

    (1) (–2,3)   (3) (2,–3)
    (2) (3,–2)   (4) (–3,2)

2. A system of equations is graphed on the set of axes below.

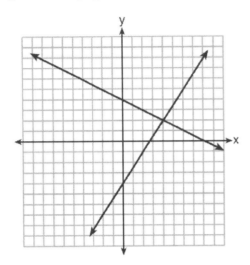

    The solution of this system is

    (1) (0,4)   (3) (4,2)
    (2) (2,4)   (4) (8,0)

3. What is the solution of the system of equations shown in the graph below?

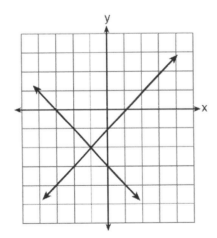

    (1) (1,0) and (–3,0)   (3) (–1,–2)
    (2) (0,–3) and (0,–1)   (4) (–2,–1)

Constructed Response

4. On the grid below, solve the system of equations graphically for *x* and *y*.

$$4x - 2y = 10$$
$$y = -2x - 1$$

5. On the set of axes below, solve the following system of equations graphically. State the coordinates of the solution.

$$y = 4x - 1$$
$$2x + y = 5$$

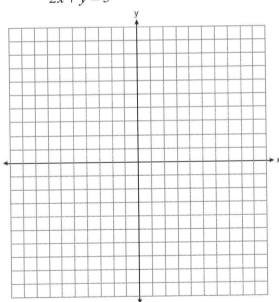

Solving Systems of Inequalities Graphically

Key Terms and Concepts

To solve a system of inequalities, graph both inequalities on the same coordinate plane. The graph of each inequality will have a shaded region representing the solution set for that inequality. The region where the two shaded regions overlap (the region that is shaded twice), including any points on a solid line bordering the region, represents the solution set of the system. Any point in the solution set would solve both inequalities.

Model Problem

Graph the following system of inequalities and label the solution set S:

$$y + 2 > 3x$$

$$-2y \geq x - 2$$

Solution:

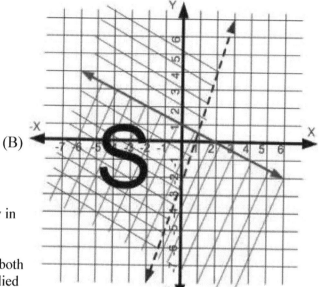

(A) $y + 2 > 3x$ $-2y \geq x - 2$

       $\underline{ -2 -2}$ $\underline{-2 -2}$

       $y > 3x - 2$ $y \leq -\frac{1}{2}x + 1$

(B)

Explanation of Steps:

   (A) If either inequality is not already in slope-intercept form, solve the inequality for y. Remember to reverse the inequality symbol if both sides of an inequality are multiplied or divided by a negative value.
     [In the second inequality, $\geq$ becomes $\leq$.]

   (B) Graph both inequalities on the same set of axes. Remember to use a solid line for inequalities with $\leq$ or $\geq$ symbols but a dashed line for those with $<$ or $>$ symbols. Label the double-shaded region "S" to represent the solution set.

Practice Problems

1. Graph the following system of inequalities and label the solution set S: $y \leq -x + 2$ $y < -1$ 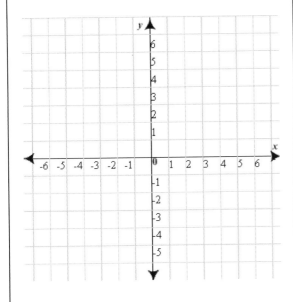	2. Graph the following system of inequalities and label the solution set S: $y \geq 2x + 1$ $y \leq -x + 4$ 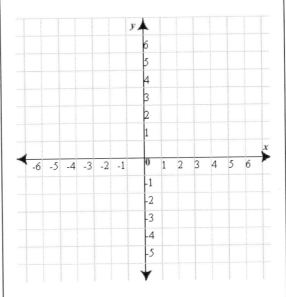
3. Graph the following system of inequalities and label the solution set S: $y < x - 2$ $2x + y \geq 1$ 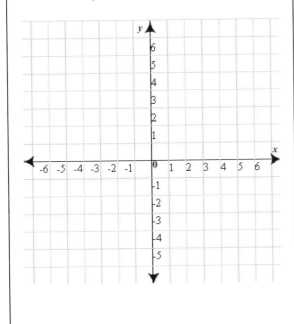	4. Graph the following system of inequalities and label the solution set S: $2x + y \geq 3$ $x - 3y < -6$ 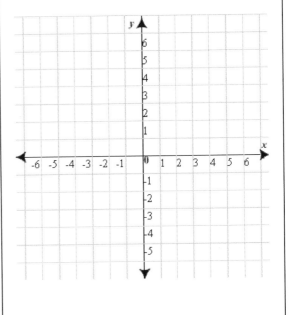

REGENTS QUESTIONS

Multiple Choice

1. Given:

$$y + x > 2$$
$$y \leq 3x - 2$$

Which graph shows the solution of the given set of inequalities?

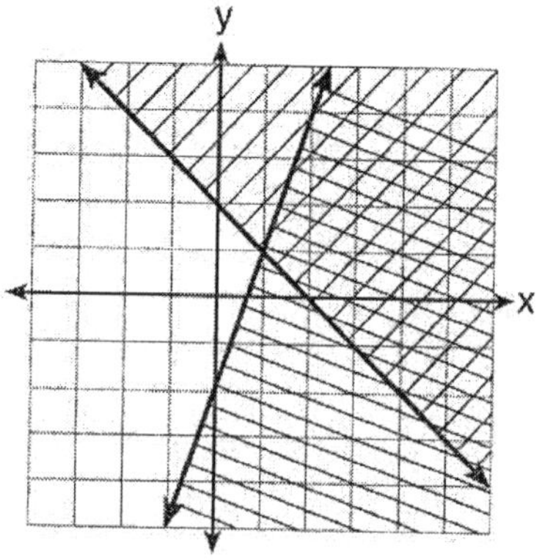

(1)

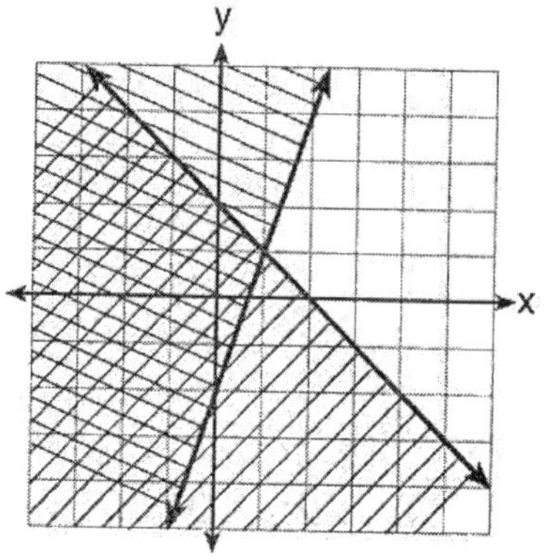

(3)

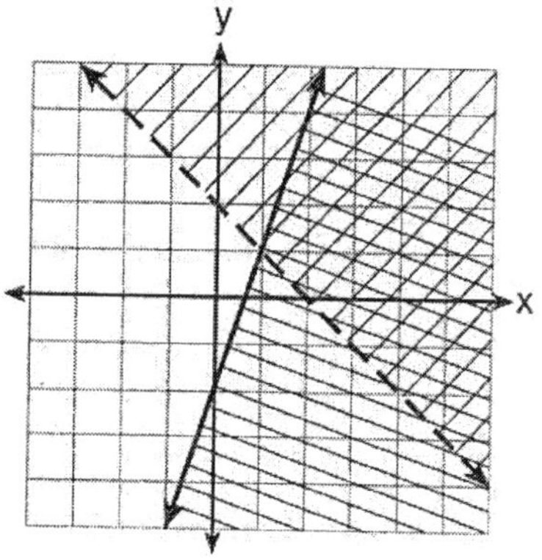

(2)

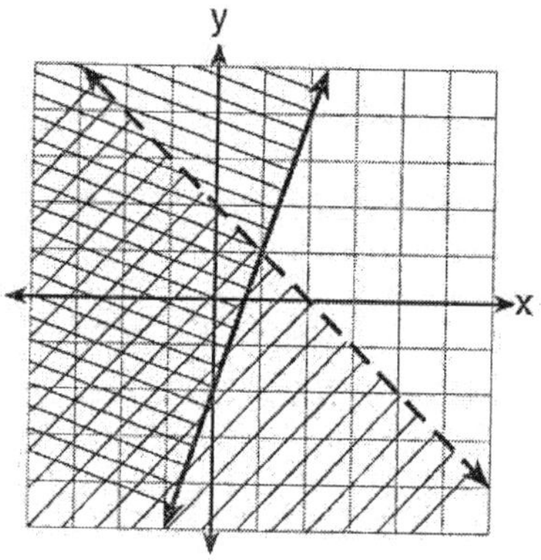

(4)

Constructed Response

2. On the set of coordinate axes below, graph each of the inequalities in the following system and label the solution set A:

$$y \leq -x + 3$$
$$y > x - 4$$

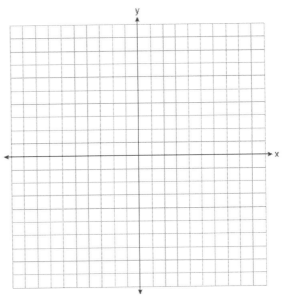

3. On the set of axes below, graph the following system of inequalities and state the coordinates of a point in the solution set.

$$2x - y \geq 6$$
$$x > 2$$

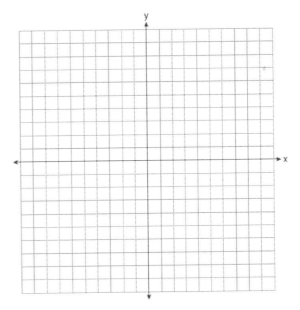

4. On the set of axes below, solve the following system of inequalities graphically.

$$y < 2x + 1$$
$$y \geq -\frac{1}{3}x + 4$$

State the coordinates of a point in the solution set.

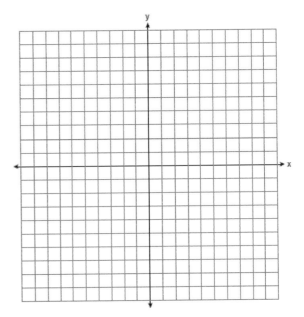

5. Graph the following systems of inequalities on the set of axes shown below and label the solution set S:

$$y > -x + 2$$
$$y \leq \frac{2}{3}x + 5$$

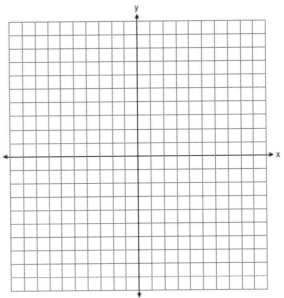

6. Solve the following system of inequalities graphically on the set of axes below.

$$3x + y < 7$$

$$y \geq \frac{2}{3}x - 4$$

State the coordinates of a point in the solution set.

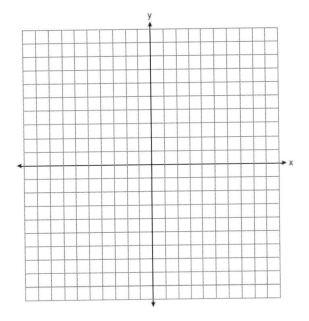

7. On the set of axes below, graph the following system of inequalities.

$$y + x \geq 3$$

$$5x - 2y > 10$$

State the coordinates of *one* point that satisfies $y + x \geq 3$, but does *not* satisfy $5x - 2y > 10$.

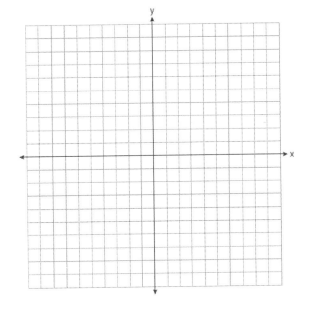

8. On the set of axes below, solve the following system of inequalities graphically.
    Label the solution set S.

$$2x + 3y < -3$$
$$y - 4x \geq 2$$

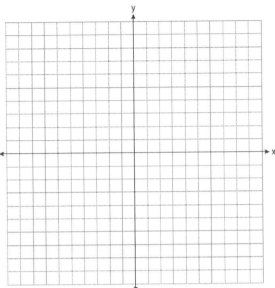

Solution Sets of Systems of Inequalities

Key Terms and Concepts

Given a graph of a system of two inequalities, a point on the graph is considered a member of the solution set for the system (a) if the point is in the **double-shaded or darker shaded (overlapping) region** of the graph, or (b) if a **solid line borders** the region, the point is on the part of the line that borders the region. Otherwise, the point is not in the solution set.

If given a system of two inequalities, you may also determine whether an ordered pair is in the solution set **without graphing**. Simply substitute the *x*- and *y*-coordinates of the ordered pair for the variables, *x* and *y*, in both inequalities. Then check if both inequalities are true.

Model Problem 1: *graphically*

Which ordered pair is in the solution set of the system of linear inequalities graphed at right?

     (1) (5,2)   (3) (−2,5)
     (2) (−5,−2)   (4) (2,−5)

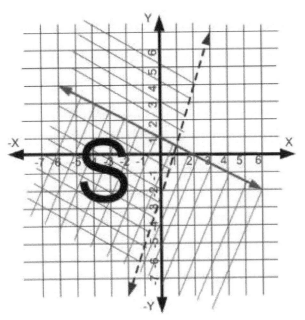

Solution:
The answer is (2). The ordered pair (−5,−2) is in the solution set.

Explanation of Steps:
Plot each point to determine whether it lies within the double-shaded region.
[(5,2) is in the unshaded region. Both (−2,5) and (2,−5) are in single-shaded regions.]

Practice Problems

1. Is (5,1) in the solution set of the system of inequalities graphed below?	2. Is (−1,−8) in the solution set of the system of inequalities graphed below?
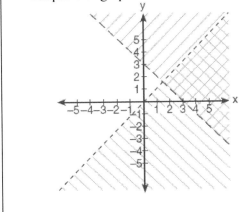	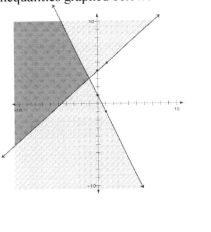

Model Problem 2: *substitution*

Which ordered pair is in the solution set of the following system of linear inequalities?

$$y + 2 > 3x$$
$$-2y \geq x - 2$$

(1) (5,2)   (2) (–5,–2)   (3) (–2,5)   (4) (2,–5)

Solution:

(A)

(5,2)	(–5,–2)	(–2,5)	(2,–5)
$y + 2 > 3x$ $(2) + 2 > 3(5)?$ $4 > 15?$ *false*	$y + 2 > 3x$ $(-2) + 2 > 3(-5)?$ $0 > -15?$ *true* $-2y \geq x - 2$ $-2(-2) \geq (-5) - 2?$ $4 \geq -7?$ *true*	$y + 2 > 3x$ $(5) + 2 > 3(-2)?$ $7 > -6?$ *true* $-2y \geq x - 2$ $-2(5) \geq (-2) - 2?$ $-10 \geq -4$ *false*	$y + 2 > 3x$ $(-5) + 2 > 3(2)?$ $-3 > 6?$ *false*
NO	YES	NO	NO

(B) The ordered pair (–5,–2) is a solution.

Explanation of Steps:

(A) For each ordered pair, substitute the coordinates for the variables *x* and *y* in the inequalities. Check whether each inequality is true for these values. If either is *false*, the ordered pair is not a solution and you may go on to check the next ordered pair.

(B) The ordered pair is in the solution set if it makes both inequalities *true*.

Checking can be done using a calculator instead. For example, to check if (5,2) is a solution:

5 [STO▸] [ALPHA] [X] [ENTER]   2 [STO▸] [ALPHA] [Y] [ENTER]

[ALPHA] [Y] + 2 [2nd] [TEST] [>] 3 [ALPHA] [X] [ENTER]   Result is [0] so (5,2) is *not* a solution.

Practice Problems

3. Which ordered pair is in the solution set of the following system of linear inequalities? $$y < 2x + 1$$ $$y \leq -3x + 4$$ (a) (1,1) (c) (1,3) (b) (1,2) (d) none of these	4. Which ordered pair is in the solution set of the following system of linear inequalities? $$y \geq x + 7$$ $$2x + y \leq -5$$ (a) (0,9) (c) (9,0) (b) (0,–9) (d) (–9,0)

REGENTS QUESTIONS

Multiple Choice

1. Which ordered pair is in the solution set of the following system of inequalities?

    $$y < \frac{1}{2}x + 4$$

    $$y \geq -x + 1$$

    (1) (-5, 3)   (3) (3, -5)
    (2) (0, 4)   (4) (4, 0)

2. Which ordered pair is in the solution set of the following system of linear inequalities?

    $$y < 2x + 2$$

    $$y \geq -x - 1$$

    (1) (0,3)   (3) (–1,0)
    (2) (2,0)   (4) (–1,–4)

3. Which ordered pair is in the solution set of the system of linear inequalities graphed below?

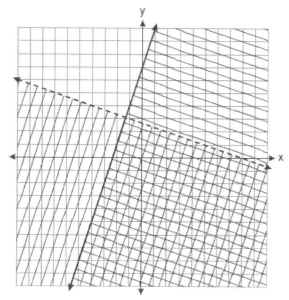

    (1) (1,–4)   (3) (5,3)
    (2) (–5,7)   (4) (–7,–2)

131

4. Which ordered pair is in the solution set of the system of inequalities shown in the graph below?

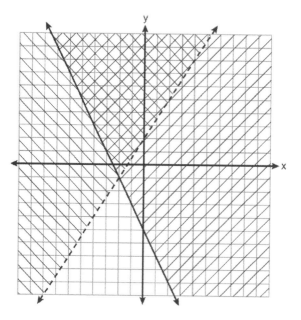

(1) (–2,–1)　　　　　　　　　(3) (–2,–4)

(2) (–2,2)　　　　　　　　　(4) (2,–2)

5. Which coordinates represent a point in the solution set of the system of inequalities shown below?

$$y \le \frac{1}{2}x + 13$$

$$4x + 2y > 3$$

(1) (–4,1)　　　　　　　　　(3) (1,–4)

(2) (–2,2)　　　　　　　　　(4) (2,–2)

6. Which ordered pair is in the solution set of the system of inequalities $y \le 3x + 1$ and $x - y > 1$?

(1) (–1,–2)　　　　　　　　　(3) (1,2)

(2) (2,–1)　　　　　　　　　(4) (–1,2)

Word Problems – Systems of Linear Equations

Key Terms and Concepts

Sometimes, the circumstances described by a word problem may lead more naturally to the use of **two variables** instead of one. If this is the case, we can find a solution by writing **two equations** in terms of the variables. We can solve for the two variables by using one of the methods learned – addition or substitution – for solving a system of equations.

Model Problem

Brenda's school is selling tickets to a spring musical. On the first day of ticket sales the school sold 3 senior citizen tickets and 9 child tickets for a total of $75. On the second day the school sold 8 senior citizen tickets and 5 child tickets for a total of $67. What is the price of each senior citizen ticket and each child ticket?

Solution:

(A) Let s represent the cost of one *senior citizen* ticket.
    Let c represent the cost of one *child* ticket.

(B)   (C)   (D)

$$3s + 9c = 75 \quad \times(8)$$
$$8s + 5c = 67 \quad \times(-3)$$

$$24s + 72c = 600$$
$$\underline{-24s - 15c = -201}$$
$$\frac{57c}{57} = \frac{399}{57}$$
$$c = 7$$

$$3s + 9(7) = 75$$
$$3s + 63 = 75$$
$$3s = 12$$
$$s = 4$$

(E) Senior citizen tickets cost $4 each and child tickets cost $7 each.

Explanation of Steps:

(A) Write what each variable represents.

(B) Write equations using the given information. *[The first day's sales lead to the first equation and the second day's sales lead to the second equation.]*

(C) Solve the system of equations for one of the variables using either the addition method or the substitution method. *[Using the addition method, we need to create additive inverses. To make additive inverses of the s terms, we can multiply the first equation by 8 and the second equation by –3. Then, adding the equations eliminates the s terms, allowing us to solve the resulting equation for c.]*

(D) Substitute the value of one variable into either original equation in order to solve for the other variable. *[Substituting 7 for c in the first original equation allows us to solve for s.]*

(E) Write your answer to the specific question posed by the word problem.

133

Practice Problems

1. The difference of two numbers is 5. Their sum is 59. Find the numbers.	2. Alexandra purchases two doughnuts and three cookies at a doughnut shop and is charged $3.30. Briana purchases five doughnuts and two cookies at the same shop for $4.95. Find the cost of one doughnut and find the cost of one cookie.
3. Tanisha and Rachel had lunch at the mall. Tanisha ordered three slices of pizza and two colas. Rachel ordered two slices of pizza and three colas. Tanisha's bill was $6.00, and Rachel's bill was $5.25. What was the price of one slice of pizza? What was the price of one cola?	4. Ramón rented a sprayer and a generator. On his first job, he used each piece of equipment for 6 hours at a total cost of $90. On his second job, he used the sprayer for 4 hours and the generator for 8 hours at a total cost of $100. What was the hourly cost of *each* piece of equipment?
5. Kristin spent $131 on shirts. Fancy shirts cost $28 and plain shirts cost $15. If she bought a total of 7 shirts then how many of each kind did she buy?	6. The sum of the digits of a certain two-digit number is 7. Reversing its digits increases the number by 9. What is the number?

REGENTS QUESTIONS

Multiple Choice

1. The equations $5x + 2y = 48$ and $3x + 2y = 32$ represent the money collected from school concert ticket sales during two class periods. If x represents the cost for each adult ticket and y represents the cost for each student ticket, what is the cost for each adult ticket?
   (1) $20   (3) $8
   (2) $10   (4) $4

2. Jack bought 3 slices of cheese pizza and 4 slices of mushroom pizza for a total cost of $12.50. Grace bought 3 slices of cheese pizza and 2 slices of mushroom pizza for a total cost of $8.50. What is the cost of one slice of mushroom pizza?
   (1) $1.50   (3) $3.00
   (2) $2.00   (4) $3.50

3. The sum of two numbers is 47, and their difference is 15. What is the larger number?
   (1) 16   (3) 32
   (2) 31   (4) 36

4. Julia went to the movies and bought one jumbo popcorn and two chocolate chip cookies for $5.00. Marvin went to the same movie and bought one jumbo popcorn and four chocolate chip cookies for $6.00. How much does one chocolate chip cookie cost?
   (1) $0.50   (3) $1.00
   (2) $0.75   (4) $2.00

Constructed Response

5. A school club held a bake sale at which a total of 40 cakes and pies were sold. The cakes sold for 90¢ each and the pies sold for 75¢ each. The club received $32.25 from the sale. How many cakes were sold?

6. One month a school store sold 15 pennants and 10 shirts for a total of $60. The next month it sold 25 pennants and 20 shirts for a total of $110. What was the selling price, in dollars, of one pennant?

7. Phil is three times as old as Carrie. Five years ago, Phil was four times as old as Carrie was at that time. How old is Carrie now?

8. The cost of 3 markers and 2 pencils is $1.80. The cost of 4 markers and 6 pencils is $2.90. What is the cost of each item? Include appropriate units in your answer.

9. The cost of three notebooks and four pencils is $8.50. The cost of five notebooks and eight pencils is $14.50. Determine the cost of one notebook and the cost of one pencil.

10. During its first week of business, a market sold a total of 108 apples and oranges. The second week, five times the number of apples and three times the number of oranges were sold. A total of 452 apples and oranges were sold during the second week. Determine how many apples and how many oranges were sold the *first* week.

11. An animal shelter spends $2.35 per day to care for each cat and $5.50 per day to care for each dog. Pat noticed that the shelter spent $89.50 caring for cats and dogs on Wednesday.

Write an equation to represent the possible numbers of cats and dogs that could have been at the shelter on Wednesday.

Pat said that there might have been 8 cats and 14 dogs at the shelter on Wednesday. Are Pat's numbers possible? Use your equation to justify your answer.

Later, Pat found a record showing that there were a total of 22 cats and dogs at the shelter on Wednesday. How many cats were at the shelter on Wednesday?

Word Problems – Systems of Inequalities

Key Terms and Concepts

There are times when a verbal problem requires more than one variable in one inequality. If the situation calls for it, use two variables and set up a system of two inequalitites. Be sure to label exactly what each variable represents.

For example: Don't write p = pencil. Instead, write p = number of pencils, or p = cost of a pencil in cents, or p = weight of a pencil in ounces, depending on the problem.

Model Problem

A home-based company produces both hand-knitted scarves and sweaters. The scarves take 2 hours of labor to produce, and the sweaters take 14 hours. The labor available is limited to 40 hours per week, and the total production capacity is 5 items per week. Write a system of inequalities representing this situation, where x is the number of scarves and y is the number of sweaters.

Solution:

(A) x = number of scarves
     y = number of sweaters

(B) $2x + 14y \le 40$

(C) $x + y \le 5$

Explanation of steps:

(A) Clearly label the variables used.

(B) Write the first inequality based on given information *[constraint on total hours]*.

(C) Write the second inequality based on given information *[constraint on number of items]*.

Practice Problems

1. You can work at most 20 hours next week. You need to earn at least $92 to cover you weekly expenses. Your dog- walking job pays $7.50 per hour and your job as a car wash attendant pays $6 per hour. Write a system of linear inequalities to model the situation.	2. Marsha is buying plants and soil for her garden. The soil cost $4 per bag, and the plants cost $10 each. She wants to buy at least 5 plants and can spend no more than $100. Write a system of linear inequalities to model the situation.
3. John is packing books into boxes. Each box can hold either 15 small books or 8 large books. He needs to pack at least 35 boxes and at least 350 books. Write a system of linear inequalities to model the situation.	4. During a family trip, you share the driving with your dad. At most, you are allowed to drive for three hours. While driving, your maximum speed is 55 miles per hour. a) Write a system of inequalities describing the possible numbers of hours t and distance d you may have to drive. b) Is it possible for you to have driven 160 miles?

137

REGENTS QUESTIONS

Constructed Response

1. A high school drama club is putting on their annual theater production. There is a maximum of 800 tickets for the show. The costs of the tickets are $6 before the day of the show and $9 on the day of the show. To meet the expenses of the show, the club must sell at least $5,000 worth of tickets.
    a) Write a system of inequalities that represent this situation.
    b) The club sells 440 tickets before the day of the show. Is it possible to sell enough additional tickets on the day of the show to at least meet the expenses of the show? Justify your answer.

2. A company manufactures bicycles and skateboards. The company's daily production of bicycles cannot exceed 10, and its daily production of skateboards must be less than or equal to 12. The combined number of bicycles and skateboards cannot be more than 16. If x is the number of bicycles and y is the number of skateboards, graph on the accompanying set of axes the region that contains the number of bicycles and skateboards the company can manufacture daily.

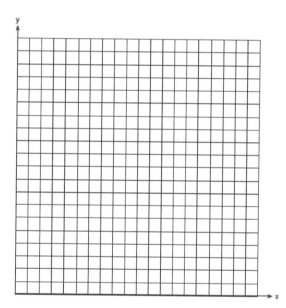

VI. POLYNOMIALS

Adding and Subtracting Polynomials

Key Terms and Concepts

A **term** is a number, a variable, or any product or quotient of numbers and variables.
A **monomial** is a single term without any variables in the denominator, such as $-2xy$.
A **polynomial** is a monomial or a sum of monomials.
For example: $x^2 + y - 2xy$ is a polynomial with 3 terms: x^2, y, and $-2xy$.
A **binomial** is a polynomial of 2 unlike terms, and a **trinomial** is a polynomial of 3 unlike terms.

A **constant term** is a term with no variable parts.

For example: In the expression $2x^2 - 3 + x + 5$, the constant terms are -3 and 5.

The **degree of a term** is the sum of its variables' exponents. A constant term has a degree of 0.

For examples: $8x^2$ has a degree of 2, $4x^3$ has a degree of 3, and $2x$ has a degree of 1.
The term $12x^2 y^3 z$ has a degree of 6 (2+3+1).

The **degree of a polynomial** is the largest degree of its terms.

For example: The degree of $9x^2 + x^4 - x$ is 4, since x^4 has the highest degree of 4.

To write a polynomial in standard form:
    (a) Combine all like terms (*simplify*)
    (b) Write terms in descending order (*exponents of a variable decrease*)

For example: $2a - 3a^2 + 4 + 9a$ $\rightarrow$ $11a - 3a^2 + 4$ $\rightarrow$ $-3a^2 + 11a + 4$

The **leading coefficient** of a polynomial is the coefficient of the term with the highest degree.
When written in standard form, this is the coefficient of the first term of the polynomial.

For example: The leading coefficient of $-3a^2 + 11a + 4$ is -3.

Adding Polynomials: Join the polynomials and simplify into standard form.
For example: $(x^2 + 5x - 24) + (2x^2 + 10) =$
                        $x^2 + 5x - 24 + 2x^2 + 10 =$
                        $3x^2 + 5x - 14$

Subtracting Polynomials: Negate all the signs of the second polynomial and then add.
For example: $(5x + 4) - (6x - 5) =$ $5x + 4 - 6x + 5 =$ $-x + 9$

When we add or subtract polynomials, the result is always a polynomial. Therefore, we can say that the set of polynomials is **closed** under addition and subtraction.

Model Problem

Subtract $-4x^2 - 12x + 5$ from $x^2 + 9x - 5$ and write the result in standard form.

Solution:

(A) $(x^2 + 9x - 5) - (-4x^2 - 12x + 5) =$

(B) $x^2 + 9x - 5 + 4x^2 + 12x - 5 =$

(C) $5x^2 + 21x - 10$

Explanation of Steps:

(A) Set up the problem. *[Remember the rule: "Subtract x from y" means y − x.]*

(B) In subtraction, negate all the terms in the second polynomial and then add.

(C) Combine like terms and express in standard form.

Practice Problems

1. Write the sum of $8x^2 - x + 4$ and $x - 5$ in standard form.	2. Write the sum of $3x^2 + x + 8$ and $x^2 - 9$ in standard form.
3. Write $(3x^2 + 2xy + 7) - (6x^2 - 4xy + 3)$ in standard form.	4. What is the result when $3a^2 - 2a + 5$ is subtracted from $a^2 + a - 1$?
5. What is the difference when $2x^2 - x + 6$ is subtracted from $x^2 - 3x - 2$?	6. Subtract $3x^2 + 4x - 1$ from $x^2 + 1$ and write the result in standard form.

REGENTS QUESTIONS

Multiple Choice

1. The expression $(2x^2 + 6x + 5) - (6x^2 + 3x + 5)$ is equivalent to

    (1) $-4x^2 + 3x$ (3) $-4x^2 - 3x + 10$

    (2) $4x^2 - 3x$ (4) $4x^2 + 3x - 10$

2. What is the sum of $x^2 - 3x + 7$ and $3x^2 + 5x - 9$?

    (1) $4x^2 - 8x + 2$ (3) $4x^2 - 2x - 2$

    (2) $4x^2 + 2x + 16$ (4) $4x^2 + 2x - 2$

3. What is the sum of $2m^2 + 3m - 4$ and $m^2 - 3m - 2$?

    (1) $m^2 - 6$ (3) $3m^2 + 6m - 6$

    (2) $3m^2 - 6$ (4) $m^2 + 6m - 2$

4. When $3g^2 - 4g + 2$ is subtracted from $7g^2 + 5g - 1$, the difference is

    (1) $-4g^2 - 9g + 3$ (3) $4g^2 + 9g - 3$

    (2) $4g^2 + g + 1$ (4) $10g^2 + g + 1$

5. When $4x^2 + 7x - 5$ is subtracted from $9x^2 - 2x + 3$, the result is

    (1) $5x^2 + 5x - 2$ (3) $-5x^2 + 5x - 2$

    (2) $5x^2 - 9x + 8$ (4) $-5x^2 + 9x - 8$

6. The sum of $4x^3 + 6x^2 + 2x - 3$ and $3x^3 + 3x^2 - 5x - 5$ is

    (1) $7x^3 + 3x^2 - 3x - 8$ (3) $7x^3 + 9x^2 - 3x - 8$

    (2) $7x^3 + 3x^2 + 7x + 2$ (4) $7x^6 + 9x^4 - 3x^2 - 8$

7. What is the result when $2x^2 + 3xy - 6$ is subtracted from $x^2 - 7xy + 2$?

    (1) $-x^2 - 10xy + 8$ (3) $-x^2 - 4xy - 4$

    (2) $x^2 + 10xy - 8$ (4) $x^2 - 4xy - 4$

8. When $5x + 4y$ is subtracted from $5x - 4y$, the difference is

    (1) 0 (3) $8y$

    (2) $10x$ (4) $-8y$

9. What is the sum of $-3x^2 - 7x + 9$ and $-5x^2 + 6x - 4$?

    (1) $-8x^2 - x + 5$ (3) $-8x^2 - 13x + 13$

    (2) $-8x^4 - x + 5$ (4) $-8x^4 - 13x^2 + 13$

141

10. When $8x^2 + 3x + 2$ is subtracted from $9x^2 - 3x - 4$, the result is
  (1) $x^2 - 2$ (3) $-x^2 + 6x + 6$
  (2) $17x^2 - 2$ (4) $x^2 - 6x - 6$

11. The sum of $3x^2 + 5x - 6$ and $-x^2 + 3x + 9$ is
  (1) $2x^2 + 8x - 15$ (3) $2x^4 + 8x^2 + 3$
  (2) $2x^2 + 8x + 3$ (4) $4x^2 + 2x - 15$

12. When $x^2 + 3x - 4$ is subtracted from $x^3 + 3x^2 - 2x$, the difference is
  (1) $x^3 + 2x^2 - 5x + 4$ (3) $-x^3 + 4x^2 + x - 4$
  (2) $x^3 + 2x^2 + x - 4$ (4) $-x^3 - 2x^2 + 5x + 4$

13. When $2x^2 - 3x + 2$ is subtracted from $4x^2 - 5x + 2$, the result is
  (1) $2x^2 - 2x$ (3) $-2x^2 - 8x + 4$
  (2) $-2x^2 + 2x$ (4) $2x^2 - 8x + 4$

14. The sum of $8n^2 - 3n + 10$ and $-3n^2 - 6n - 7$ is
  (1) $5n^2 - 9n + 3$ (3) $-11n^2 - 9n - 17$
  (2) $5n^2 - 3n - 17$ (4) $-11n^2 - 3n + 3$

15. If $A = 3x^2 + 5x - 6$ and $B = -2x^2 - 6x + 7$, then $A - B$ equals
  (1) $-5x^2 - 11x + 13$ (3) $-5x^2 - x + 1$
  (2) $5x^2 + 11x - 13$ (4) $5x^2 - x + 1$

16. What is the result when $4x^2 - 17x + 36$ is subtracted from $2x^2 - 5x + 25$?
  (1) $6x^2 - 22x + 61$ (3) $-2x^2 - 22x + 61$
  (2) $2x^2 - 12x + 11$ (4) $-2x^2 + 12x - 11$

17. When $6x^2 - 4x + 3$ is subtracted from $3x^2 - 2x + 3$, the result is
  (1) $3x^2 - 2x$ (3) $3x^2 - 6x + 6$
  (2) $-3x^2 + 2x$ (4) $-3x^2 - 6x + 6$

Constructed Response

18. From $4x^2 - 8x$ subtract $7x^2 + 3x$.

19. Express as a binomial in terms of x the perimeter of a triangle whose sides are represented by $4x - 8$, $6x + 3$, and $x + 1$.

20. Subtract $5x^2 - 7x - 6$ from $9x^2 + 3x - 4$.

21. Subtract $2x^2 - 5x + 8$ from $6x^2 + 3x - 2$ and express the answer as a trinomial.

Multiplying Polynomials

Key Terms and Concepts

Factors are any parts of an expression that are multiplied to produce a product.

For examples: (a) In $2 \cdot 3$, both 2 and 3 are factors.

(b) In $2x^2 y$, the factors are 2, x^2 and y.

(c) In $3(x+1)$, both 3 and $(x+1)$ are factors.

(d) The expression $(a-2)(a+3)$ has two binomial factors.

Multiplying a Monomial by a Polynomial: Apply the Distributive Property.

For examples: (a) $3(x^2 + x)$ $\rightarrow$ $3x^2 + 3x$

(b) $-(2m^4 - n^2)$ $\rightarrow$ $-2m^4 + n^2$

(c) $5a(a^3 - 3a + 1)$ $\rightarrow$ $5a^4 - 15a^2 + 5a$

Multiplying Binomials

Method 1: Distribution – Multiply term 1 by the second binomial, then term 2 by the second binomial

$$(a-2)(a+3) = a(a+3) - 2(a+3) =$$
$$a^2 + 3a - 2a - 6 =$$
$$a^2 + a - 6$$

Method 2: Rectangle diagram – Write the terms of each factor outside a side of a rectangle, find the algebraic areas of each inner rectangle, and add the inner areas

$$(a-2)(a+3)$$

	a	-2
a	a^2	$-2a$
3	$3a$	-6

$$a^2 + 3a - 2a - 6 =$$
$$a^2 + a - 6$$

Method 3: "FOIL" – Multiply these pairs of terms: Firsts, Outers, Inners, Lasts

$$(a-2)(a+3) =$$

$\quad F \quad O \quad I \quad L$

$$a^2 + 3a - 2a - 6 =$$
$$a^2 + a - 6$$

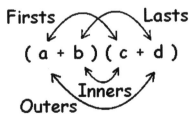

Squaring a Binomial: multiply the binomial by itself.

For example: $(x-3)^2 = (x-3)(x-3) = x^2-3x-3x+9 = x^2-6x+9$

Multiplying a Binomial by a Trinomial

Expand either Method 1 (Distribution) or Method 2 (Rectangle diagram) above. Note that the "FOIL" method only works for multiplying two binomials.

Method 1: Distribution – Multiply term 1 by the trinomial, then term 2 by the trinomial.

$$(a+2)(2a^2-5a+3) = a(2a^2-5a+3)+2(2a^2-5a+3) =$$
$$2a^3-5a^2+3a \ + \ 4a^2-10a+6 =$$
$$2a^3-a^2-7a+6$$

Method 2: Rectangle diagram – Write the terms of each factor outside a side of a rectangle, find the algebraic areas of each inner rectangle, and add the inner areas

$$(a+2)(2a^2-5a+3)$$

	a	2
$2a^2$	$2a^3$	$4a^2$
$-5a$	$-5a^2$	$-10a$
3	$3a$	6

$$2a^3-5a^2+3a \ + \ 4a^2-10a+6 =$$
$$2a^3-a^2-7a+6$$

When we multiply polynomials, the result is always a polynomial. Therefore, we can say that the set of polynomials is **closed** under multiplication. This is not true when we divide polynomials, since we may end up with an algebraic fraction (rational expression) that is not a polynomial.

Model Problem

Write the product of $x+8$ and $x-2$ as a polynomial in standard form.

Solution:

(A) $(x+8)(x-2) =$

(B) $x^2-2x+8x-16 =$

(C) $x^2+6x-16$

Explanation of Steps:

(A) Set up the problem.

(B) Use the FOIL method. *[Firsts = $x \cdot x$, Outers = $x \cdot (-2)$, Inners = $8 \cdot x$, Lasts = $8 \cdot (-2)$.]*

(C) Combine like terms and express in standard form.

Practice Problems

1. Multiply: $7x(1-x^3)$	2. What is the product of $2r^2 - 5$ and $3r$?
3. What is the product of $(c+8)$ and $(c-5)$?	4. Express $(x-7)(2x+3)$ as a trinomial.
5. Use the following diagram to expand $(a+b)^2$ into an equivalent polynomial.	6. Simplify the expression $(x-6)^2$.
7. Create a rectangle diagram for $(x+3)(x-y-1)$.	8. Which of the following expressions is *not* equivalent to $(a+b)(x+y)$? (a) $(a+b)x+(a+b)y$ (b) $a(x+y)+b(x+y)$ (c) $ax+by$ (d) $ax+bx+ay+by$
9. Multiply $(x-1)(2x^2+x-2)$ and write the product in standard form.	10. Multiply $(x^2+2)(x^2-2x+1)$ and write the product in standard form.

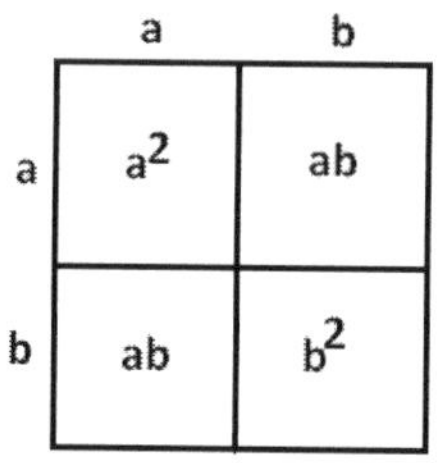

REGENTS QUESTIONS

Multiple Choice

1. If $(2a+5)$ is squared, the result is

    (1) $2a^2 + 25$ (3) $4a^2 + 10a + 25$

    (2) $4a^2 + 25$ (4) $4a^2 + 20a + 25$

2. The expression $x(x-y)(x+y)$ is equivalent to

    (1) $x^2 - y^2$ (3) $x^3 - xy^2$

    (2) $x^3 - y^3$ (4) $x^3 - x^2y + y^2$

3. The expression $(2x+1)^2 - 2(2x^2 - 1)$ is equivalent to

    (1) $4x + 3$ (3) 3

    (2) $2x + 3$ (4) -1

4. What is the product of $-3x^2y$ and $(5xy^2 + xy)$?

    (1) $-15x^3y^3 - 3x^3y^2$ (3) $-15x^2y^2 - 3x^2y$

    (2) $-15x^3y^3 - 3x^3y$ (4) $-15x^3y^3 + xy$

5. The length of a rectangular room is 7 less than three times the width, w, of the room. Which expression represents the area of the room?

    (1) $3w - 4$ (3) $3w^2 - 4w$

    (2) $3w - 7$ (4) $3w^2 - 7w$

6. Which expression is equivalent to $-3x(x-4) - 2x(x+3)$?

    (1) $-x^2 - 1$ (3) $-5x^2 - 6x$

    (2) $-x^2 + 18x$ (4) $-5x^2 + 6x$

7. What is the product of $(3x+2)$ and $(x-7)$?

    (1) $3x^2 - 14$ (3) $3x^2 - 19x - 14$

    (2) $3x^2 - 5x - 14$ (4) $3x^2 - 23x - 14$

Constructed Response

8. Express the product $(1+x)(1-x)(1+x^2)$ as a binomial.

9. Express as a trinomial the product of $x+1$ and $x-3$.

Dividing a Polynomial by a Monomial

Key Terms and Concepts

When **dividing a polynomial by a monomial**, we may be able to simplify by dividing each term of the polynomial in the numerator by the monomial in the denominator. For each new fraction, divide the coefficients and subtract exponents of the same base. The result should have as many terms as the original polynomial.

For example: $\dfrac{21a^2b - 3ab}{3ab} = \dfrac{21a^2b}{3ab} - \dfrac{3ab}{3ab} = 7a - 1$

Note: In Algebra II, we will learn how to divide by a polynomial of more than one term.

Practice Problems

1. Divide $\dfrac{2x + 4}{2}$	2. Write $\dfrac{x^2 + 2x}{x}$ in simplest form.
3. Divide: $\dfrac{14ab + 28b}{14b}$	4. Simplify: $\dfrac{6x^3 + 9x^2 + 3x}{3x}$

REGENTS QUESTIONS

Multiple Choice

1. Which expression represents $\dfrac{12x^3 - 6x^2 + 2x}{2x}$ in simplest form?

       (1) $6x^2 - 3x$ (3) $6x^2 - 3x + 1$

       (2) $10x^2 - 4x$ (4) $10x^2 - 4x + 1$

2. The quotient of $\dfrac{8x^5 - 2x^4 + 4x^3 - 6x^2}{2x^2}$ is

       (1) $16x^7 - 4x^6 + 8x^5 - 12x^4$ (3) $4x^3 - x^2 + 2x - 3x$

       (2) $4x^7 - x^6 + 2x^5 - 3x^4$ (4) $4x^3 - x^2 + 2x - 3$

3. Which expression is equivalent to $\dfrac{2x^6 - 18x^4 + 2x^2}{2x^2}$?

       (1) $x^3 - 9x^2$ (3) $x^3 - 9x^2 + 1$

       (2) $x^4 - 9x^2$ (4) $x^4 - 9x^2 + 1$

4. What is $24x^2y^6 - 16x^6y^2 + 4xy^2$ divided by $4xy^2$?

       (1) $6xy^4 - 4x^5$ (3) $6x^2y^3 - 4x^6y$

       (2) $6xy^4 - 4x^5 + 1$ (4) $6x^2y^3 - 4x^6y + 1$

5. When $16x^3 - 12x^2 + 4x$ is divided by $4x$, the quotient is

       (1) $12x^2 - 8x$ (3) $4x^2 - 3x$

       (2) $12x^2 - 8x + 1$ (4) $4x^2 - 3x + 1$

Constructed Response

6. Express in simplest form: $\dfrac{45a^4b^3 - 90a^3b}{15a^2b}$

VII.   RADICALS

Irrational Numbers

Key Terms and Concepts

We have seen that the set of real numbers is made up of **rational numbers** and **irrational numbers**. (In the language of sets, the set of reals is the union of the two disjoint sets, the set of rationals and the set of irrationals.)

A rational number is any number that can be expressed as a fraction $\frac{a}{b}$ where a is an integer and b is a non-zero integer. Expressed in decimal form, rational numbers are terminating or repeating decimals, such as -100, -1.75. or $2.\overline{6}$. So, an irrational number is any real number that *cannot* be expressed as a fraction of integers. Irrational numbers, such as π and $\sqrt{2}$, are non-terminating, non-repeating decimals.

In fact, the *square roots of all whole numbers that are not perfect squares are irrational*.
For examples: $\sqrt{9}$ and $\sqrt{49}$ are rational, but $\sqrt{3}$ and $\sqrt{50}$ are irrational.

$\sqrt{x}$ represents the principal square root of x, or the non-negative value that, when multiplied by itself, is equal to x. The $\sqrt{}$ symbol is called a **radical sign**, and the quantity under the radical sign is called the **radicand**. A **radical** is any term containing a radical sign.
For example: $\sqrt{36} = 6$ (36 is the radicand.)

Find the square root of a number on the calculator using the $\boxed{\text{2nd}}$ $\boxed{\sqrt{}}$ keys. The symbols "$\sqrt{}($" appear; after typing the radicand, remember to press the close parentheses $\boxed{)}$ key.
For example: Find $\sqrt{36}$ by entering $\boxed{\text{2nd}}$ $\boxed{\sqrt{}}$ 36 $\boxed{)}$ $\boxed{\text{ENTER}}$.

Since a positive number has a **negative square root** as well, we represent the negative square root by placing a negative sign before the radical. To indicate both the positive and negative roots, we use the $\pm$ **symbol** before the radical.
For examples: $\sqrt{36} = 6$ $-\sqrt{36} = -6$ since $(-6)(-6) = 36$ $\pm\sqrt{36} = \{-6, 6\}$

It is important to recognize that **squaring** and taking a **square root** are reverse operations.
Therefore, $\sqrt{x^2} = x$ and $\left(\sqrt{y}\right)^2 = y$.
For examples: $\sqrt{3^2} = \sqrt{9} = 3$ and $\left(\sqrt{25}\right)^2 = 5^2 = 25$.

Also important facts are that $\sqrt{ab} = \sqrt{a}\sqrt{b}$ and $\sqrt{\dfrac{a}{b}} = \dfrac{\sqrt{a}}{\sqrt{b}}$ (for $b \neq 0$).

For examples: $\sqrt{36} = \sqrt{9 \cdot 4} = \sqrt{9}\sqrt{4} = 3 \cdot 2 = 6$ and $\sqrt{\dfrac{4}{9}} = \dfrac{\sqrt{4}}{\sqrt{9}} = \dfrac{2}{3}$.

The set of irrational numbers is **not closed** under any of the four basic operations.

For examples: $\sqrt{2}+(-\sqrt{2})=0$, $\sqrt{2}-\sqrt{2}=0$, $\sqrt{2}\cdot\sqrt{2}=2$, and $\dfrac{\sqrt{2}}{\sqrt{2}}=1$.

When any of the four basic operations are performed on *non-zero* real numbers, if one of the operands is **rational** and the other operand is **irrational**, the result is **irrational**.

How do we know? Remember, an irrational number is a non-repeating, non-terminating decimal. If we were to, say, add such a number to a rational number (that is, a terminating or repeating decimal), the result would still be a non-repeating, non-terminating decimal.

For examples: (a) $2+\sqrt{2}=2+1.414213...=3.414213...$

(b) $\dfrac{1}{3}+\pi=0.33333...+3.14159...=3.47492...$

Another way to look at it is by using our knowledge of closure:

For example: Is it possible to have $a+b=c$ where a is rational, b is irrational, and c is rational? The answer is no, by the following proof:

1) If we solve the equation for b, we get $b=c-a$.
2) If both c and a are rational, and we know the set of rational numbers is closed under subtraction, then the difference $c-a$ would have to be rational.
3) This would mean that b would have to be rational, since $b=c-a$.
4) Since we originally said b is irrational, we've come to a contradiction.
5) So, it's not possible to have $a+b=c$ where a is rational, b is irrational, and c is rational. Therefore the sum c would have to be irrational.

(This is known in higher math as a *proof by contradiction* or an *indirect proof*.)

Similar explanations can be used to justify or prove this for the other basic operations.

For example: Let's multiply 5 by π. 5π cannot be rational, because if it were, then when we multiply 5π by the rational number $\dfrac{1}{5}$, the product would have to be rational, since the set of rational numbers is closed under multiplication. However, $5\pi\cdot\dfrac{1}{5}=\pi$ and we know π is irrational. So, 5π must also be irrational.

Practice Problems

1. Which of the following square roots is an irrational number? (1) $-\sqrt{16}$ (3) $\sqrt{64}$ (2) $\sqrt{8}$ (4) $\sqrt{\dfrac{1}{64}}$	2. State whether $2\sqrt{3}$ is rational or irrational. Justify your answer.
3. State whether $\dfrac{\pi}{2}$ is rational or irrational. Justify your answer.	4. State whether $\dfrac{2-\sqrt{29}}{4}$ is rational or irrational. Justify your answer.
5. Name at least three possible values of x that would make $x\sqrt{3}$ a rational number.	6. $\dfrac{22}{7}$ and 3.14 are often used as a rational approximations of π. Which of these is closer to the actual value of π?

REGENTS QUESTIONS

Multiple Choice

1. Given:
   $$L = \sqrt{2}$$
   $$M = 3\sqrt{3}$$
   $$N = \sqrt{16}$$
   $$P = \sqrt{9}$$

   Which expression results in a rational number?

   (1) $L + M$ (3) $N + P$

   (2) $M + N$ (4) $P + L$

Simplifying Radicals

Key Terms and Concepts

To simplify a radical into simplest radical form:
  (A) Write the prime factorization of the radicand.
  (B) Group all pairs of factors, representing squares.
  (C) Remove the squares (pairs of factors) from the radicand by replacing them with their square roots (single factors outside the radical sign).
  (D) Multiply all factors outside the radicand, and all factors remaining inside the radicand.

For examples: $\sqrt{75} = \sqrt{3 \cdot \boxed{5 \cdot 5}} = 5\sqrt{3}$

$\sqrt{288} = \sqrt{\boxed{2 \cdot 2} \cdot \boxed{2 \cdot 2} \cdot 2 \cdot \boxed{3 \cdot 3}} = 2 \cdot 2 \cdot 3 \cdot \sqrt{2} = 12\sqrt{2}$

Model Problem
Simplify $8\sqrt{90}$

Solution:

$\qquad$ (A) (B) $\qquad$ (C) $\qquad\qquad$ (D)

$8\sqrt{90} = 8\sqrt{2 \cdot \boxed{3 \cdot 3} \cdot 5} = 8 \cdot 3\sqrt{2 \cdot 5} = 24\sqrt{10}$

Explanation of Steps:
  (A) Write the prime factorization of the radicand.
  (B) Group all pairs of factors, representing squares.
  (C) Remove the squares (pairs of factors) from the radicand by replacing them with their square roots (single factors) outside the radical sign.
  (D) Multiply all factors outside the radicand, and all factors remaining inside the radicand.

Practice Problems

1. Simplify $\sqrt{12}$.	2. Simplify $\sqrt{50}$.
3. Simplify $5\sqrt{72}$.	4. Simplify $3\sqrt{45}$.
5. Simplify $2\sqrt{128}$.	6. Simplify $\dfrac{7\sqrt{18}}{3}$.

REGENTS QUESTIONS

Multiple Choice

1. What is $\dfrac{\sqrt{32}}{4}$ expressed in simplest radical form?

   (1) $\sqrt{2}$ (3) $\sqrt{8}$

   (2) $4\sqrt{2}$ (4) $\dfrac{\sqrt{8}}{2}$

2. What is $\sqrt{72}$ expressed in simplest radical form?

   (1) $2\sqrt{18}$ (3) $6\sqrt{2}$

   (2) $3\sqrt{8}$ (4) $8\sqrt{3}$

3. What is $\sqrt{32}$ expressed in simplest radical form?

   (1) $16\sqrt{2}$ (3) $4\sqrt{8}$

   (2) $4\sqrt{2}$ (4) $2\sqrt{8}$

4. When $5\sqrt{20}$ is written in simplest radical form, the result is $k\sqrt{5}$. What is the value of k?

   (1) 20   (3) 7

   (2) 10   (4) 4

5. What is $3\sqrt{250}$ expressed in simplest radical form?

   (1) $5\sqrt{10}$ (3) $15\sqrt{10}$

   (2) $8\sqrt{10}$ (4) $75\sqrt{10}$

6. What is $2\sqrt{45}$ expressed in simplest radical form?

   (1) $3\sqrt{5}$ (3) $6\sqrt{5}$

   (2) $5\sqrt{5}$ (4) $18\sqrt{5}$

Constructed Response

7. Express $5\sqrt{72}$ in simplest radical form.

8. Express $-3\sqrt{48}$ in simplest radical form.

9. Express $4\sqrt{75}$ in simplest radical form.

10. Express $2\sqrt{108}$ in simplest radical form.

Operations with Radicals

Key Terms and Concepts

Radicals may be combined by addition or subtraction only if, when expressed in simplest radical form, they are like radicals. **Like radicals** have the **same radicand**.
(*Note:* they must also have the same index, but we are only concerned with square roots here.)

Sometimes unlike radicals may be simplified into like radicals.
For example: $\sqrt{12}$ and $\sqrt{75}$ can be simplified into $2\sqrt{3}$ and $5\sqrt{3}$, which are like radicals.

Combine like radicals just as you would combine like terms: add or subtract the coefficients and keep the radicand unchanged.
For example: $2\sqrt{3}+5\sqrt{3}-\sqrt{3}=(2+5-1)\sqrt{3}=6\sqrt{3}$

To multiply radicals, separately find the product of their coefficients and the product of their radicands, then simplify if possible.
For example: $(5\sqrt{3})(2\sqrt{7})=10\sqrt{21}$

To divide radicals, separately find the quotient of their coefficients and the quotient of their radicands, then simplify if possible.
For example: $\dfrac{6\sqrt{72}}{3\sqrt{8}}=2\sqrt{9}=2\cdot3=6$

Sometimes, multiple operations involving radicals need to be performed.
For example: Simplify $\sqrt{2}(\sqrt{10}+4)$ using the distributive property.
$$\sqrt{2}(\sqrt{10}+4)=\sqrt{20}+4\sqrt{2}=2\sqrt{5}+4\sqrt{2}$$

Model Problem 1: *combining radicals by addition or subtraction*
Express the sum $3\sqrt{8}+2\sqrt{2}$ in simplest radical form.

Solution:
    (A) $3\sqrt{8}$ can be simplified as follows: $3\sqrt{\boxed{2\cdot2}\cdot2}=3\cdot2\sqrt{2}=6\sqrt{2}$.
    (B) So, $3\sqrt{8}+2\sqrt{2}=6\sqrt{2}+2\sqrt{2}=8\sqrt{2}$

Explanation of Steps:
    (A) Express each term in simplest radical form.
    (B) Combine like radicals by adding or subtracting their coefficients.

Practice Problems

1. Add $\sqrt{75} + \sqrt{3}$.	2. Add $\sqrt{27} + \sqrt{12}$.
3. Add $5\sqrt{7} + 3\sqrt{28}$.	4. Subtract $2\sqrt{50} - \sqrt{2}$.

Model Problem 2: *multiplying radicals*

Express the product $(5\sqrt{8})(7\sqrt{3})$ in simplest radical form.

Solution:

(A) $(5\sqrt{8})(7\sqrt{3}) = 35\sqrt{24}$

(B) Simplifying, we get $35\sqrt{24} = 35\sqrt{\boxed{2 \cdot 2} \cdot 2 \cdot 3} = 35 \cdot 2\sqrt{2 \cdot 3} = 70\sqrt{6}$

Explanation of Steps:

(A) Find the product of the coefficients *[5 × 7 = 35].*
   Find the product of the radicands *[8 × 3 = 24].*

(B) Express in simplest radical form.

Practice Problems

5. Express $\sqrt{6} \cdot \sqrt{15}$ in simplest form.	6. Express in simplest form: $\sqrt{90} \cdot \sqrt{40} - \sqrt{8} \cdot \sqrt{18}$

Model Problem 3: *dividing radicals*

Express $\dfrac{9\sqrt{20}}{3\sqrt{5}}$ in simplest radical form.

Solution:

$$\frac{9\sqrt{20}}{3\sqrt{5}} = 3\sqrt{4} = 3 \cdot 2 = 6$$

Explanation of Steps:

(A) Find the quotient of the coefficients *[9 ÷ 3 = 3]*.
Find the quotient of the radicands *[20 ÷ 5 = 4]*.

(B) Express in simplest radical form.

Practice Problems

7. Express $\dfrac{\sqrt{65}}{\sqrt{5}}$ in simplest form.	8. Express in simplest form: $\dfrac{20\sqrt{100}}{4\sqrt{2}}$
9. Express in simplest form: $\dfrac{\sqrt{48} - 5\sqrt{27} + 2\sqrt{75}}{\sqrt{3}}$	10. Express in simplest form: $\dfrac{\sqrt{27} + \sqrt{75}}{\sqrt{12}}$

REGENTS QUESTIONS

Multiple Choice

1. The expression $\sqrt{48} - \sqrt{12}$ is equivalent to

   (1) 6   (3) $3\sqrt{2}$

   (2) 2   (4) $2\sqrt{3}$

2. The expression $5\sqrt{2} - \sqrt{32}$ is equivalent to

   (1) $-11\sqrt{2}$ (3) $9\sqrt{2}$

   (2) $\sqrt{2}$ (4) $\sqrt{18}$

3. The expression $6\sqrt{50} + 6\sqrt{2}$ written in simplest radical form is

   (1) $6\sqrt{52}$ (3) $17\sqrt{2}$

   (2) $12\sqrt{52}$ (4) $36\sqrt{2}$

4. The expression $\sqrt{72} - 3\sqrt{2}$ written in simplest radical form is

   (1) $5\sqrt{2}$ (3) $3\sqrt{2}$

   (2) $3\sqrt{6}$ (4) $\sqrt{6}$

5. What is $3\sqrt{2} + \sqrt{8}$ expressed in simplest radical form?

   (1) $3\sqrt{10}$ (3) $5\sqrt{2}$

   (2) $3\sqrt{16}$ (4) $7\sqrt{2}$

6. The product of $(3 + \sqrt{5})$ and $(3 - \sqrt{5})$ is

   (1) $4 - 6\sqrt{5}$ (3) 14

   (2) $14 - 6\sqrt{5}$ (4) 4

Constructed Response

7. Express the product of $3\sqrt{20}(2\sqrt{5}-7)$ in simplest radical form.

8. Express $\dfrac{16\sqrt{21}}{2\sqrt{7}}-5\sqrt{12}$ in simplest radical form.

9. Express $\dfrac{3\sqrt{75}+\sqrt{27}}{3}$ in simplest radical form.

10. Express $\sqrt{25}-2\sqrt{3}+\sqrt{27}+2\sqrt{9}$ in simplest radical form.

11. Express $\dfrac{\sqrt{84}}{2\sqrt{3}}$ in simplest radical form.

12. Perform the indicated operations and express the answer in simplest radical form.

$$3\sqrt{7}\left(\sqrt{14}+4\sqrt{56}\right)$$

VIII. CLASSIFICATION OF STATISTICAL DATA

Population and Sample

Key Terms and Concepts

Statistics is the study of the collection and analysis of data. One common method of collecting data is by conducting a **survey**, in which a specific question is posed to a sample of people and their responses are recorded. Another common method is by performing an **experiment** and recording the results.

Rarely can we collect data from the entire **population** of all possible subjects. Generally, a survey of only a **sample** (usually a subset) of the entire population is taken.

For example, suppose we want to find the average age among *all registered U.S. voters*. This group represents the *population*, estimated at over 150 million people (which is still only half of the estimated general population of the U.S.). To determine the average age, we might take a *sample* of, say, one million registered voters. Even if this sample is random and unbiased, the average age of the sample may be close to, but not necessarily equal to, the average age of the population. After all, even this very large sample is still less than 1% of the entire population of registered voters. Since it is impossible to survey all registered voters at any one moment in time, we could never determine the population's exact average age from any sample.

Model Problem:
A survey of 1353 American households found that 18% of the households own a computer. Identify the population and sample.

Solution:
>   population = all American households
>   sample = the 1353 American houeholds that were surveyed

Explanation:
>   The population is the entire set we want to study. The sample is the subset that we actually survey.

Practice Problems

1. Identify the population and the sample:	2. Identify the population and the sample:
A manufacturer received a large shipment of bolts. The bolts must meet certain specifications to be useful. Before accepting shipment, 100 bolts were selected, and it was determined whether or not each met specifications.	The average weight of every sixth person entering the mall within a 3 hour period was 146 lb.

Qualitative and Quantitative Data

Key Terms and Concepts

A set of data may be qualitative or quantitative in nature:

Data is **qualitative** (from the word, "quality") if it is recorded as non-numeric characteristics. Qualitative data is also known as **categorical**.
For example: If a survey asks for a favorite soda, the responses are qualitative.

Data is **quantitative** (from the word, "quantity") if it is recorded as numeric values. Quantitative data allows for numerical analysis of the results, such as finding the average (mean) of the data. Quantitative data is also known as **numerical**.
For example: If a survey asks for test scores, the responses are quantitative.

Model Problem
In a survey, students are asked how many pages of their assigned novel they've read so far. Is this set of data qualitative or quantitative?

Solution:
Quantitative

Explanation of Steps:
If the responses are numeric values *[numbers of pages]*, the data is quantitative.

Practice Problems

1. Identify the set of data as qualitative or quantitative: The local temperatures from a number of U.S. cities at a given time.	2. Identify the set of data as qualitative or quantitative: The opinions of students regarding their school's safety.

REGENTS QUESTIONS

Multiple Choice

1. Which data set describes a situation that could be classified as qualitative?
    (1) the elevations of the five highest mountains in the world
    (2) the ages of presidents at the time of their inauguration
    (3) the opinions of students regarding school lunches
    (4) the shoe sizes of players on the basketball team

2. Which data set describes a situation that could be classified as qualitative?
    (1) the ages of the students in Ms. Marshall's Spanish class
    (2) the test scores of the students in Ms. Fitzgerald's class
    (3) the favorite ice cream flavor of each of Mr. Hayden's students
    (4) the heights of the players on the East High School basketball team

3. Which data set describes a situation that could be classified as quantitative?
    (1) the phone numbers in a telephone book
    (2) the addresses for students at Hopkins High School
    (3) the zip codes of residents in the city of Buffalo, New York
    (4) the time it takes each of Mr. Harper's students to complete a test

4. Which set of data can be classified as qualitative?
    (1) scores of students in an algebra class
    (2) ages of students in a biology class
    (3) numbers of students in history classes
    (4) eye colors of students in an economics class

5. Which set of data can be classified as quantitative?
    (1) first names of students in a chess club
    (2) ages of students in a government class
    (3) hair colors of students in a debate club
    (4) favorite sports of students in a gym class

6. Craig sees an advertisement for a car in a newspaper. Which information would *not* be classified as quantitative?
    (1) the cost of the car   (3) the model of the car
    (2) the car's mileage   (4) the weight of the car

7. Which set of data describes a situation that could be classified as qualitative?
    (1) the colors of the birds at the city zoo
    (2) the shoe size of the zookeepers at the city zoo
    (3) the heights of the giraffes at the city zoo
    (4) the weights of the monkeys at the city zoo

8. An art studio has a list of information posted with each sculpture that is for sale. Each entry in the list could be classified as quantitative *except* for the

    (1) cost   (3) artist

    (2) height   (4) weight

9. Which data can be classified as quantitative?

    (1) favorite stores at which you shop

    (2) U.S. Representatives and their home states

    (3) sales tax rate in each New York county

    (4) opinion of a freshman on the color of Paul's shirt

10. Which set of data is qualitative?

    (1) laps swum in a race

    (2) number of swimmers on the team

    (3) swimmers' favorite swimsuit colors

    (4) temperature in Fahrenheit of the water in a pool

Univariate and Bivariate Data

Key Terms and Concepts

Data is considered **univariate** if it can be recorded using a **single variable**. A frequency table or histogram uses univariate data. **Frequency** represents the number of times each result or response occurs in the data. Univariate data can also be represented by a box-and-whisker plot.

For example: A survey asks a sampling of people for their heights in inches.

Data is considered **bivariate** if it is recorded using **two variables**. Bivariate data can be represented by plotting points on a coordinate graph or scatter plot. For each result or response, an ordered pair represents the values of the two variables, usually *x* and *y*, for that point.

For example: A survey asks a sampling of people for their heights in inches and their weights in pounds.

Model Problem

30 students took a 5-point quiz and the results are shown in the table below. Identify the data as either univariate or bivariate.

Score	Frequency
0	4
1	3
2	5
3	5
4	6
5	7

Solution:

Univariate

Explanation of Steps:

Determine whether one or two variables are needed to record the responses to a survey or results of an experiment. *[The only variable here, for each of the 30 students, is for the quiz score. The frequency is simply a count of how many times each possible score occurred.]*

Practice Problems

1. The following graph shows temperatures and amounts of revenue over a 12 day period for an ice cream shop. Identify the data as univariate or bivariate.

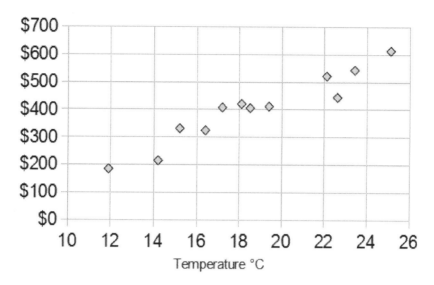

2. A certain baseball pitcher throws four types of pitches in a game: a curveball, fastball, changeup, or slider. The bar graph below shows how many of each type of pitch was thrown by the pitcher during the game. Identify the data as univariate or bivariate.

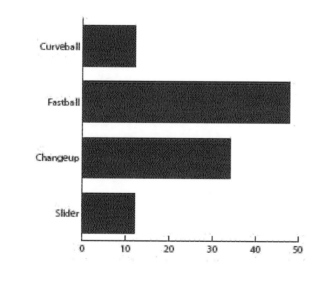

REGENTS QUESTIONS

Multiple Choice

1. Which situation should be analyzed using bivariate data?
    - (1) Ms. Saleem keeps a list of the amount of time her daughter spends on her social studies homework.
    - (2) Mr. Benjamin tries to see if his students' shoe sizes are directly related to their heights.
    - (3) Mr. DeStefan records his customers' best video game scores during the summer.
    - (4) Mr. Chan keeps track of his daughter's algebra grades for the quarter.

2. Which data table represents univariate data?

Side Length of a Square	Area of Square
2	4
3	9
4	16
5	25

(1)

Age Group	Frequency
20–29	9
30–39	7
40–49	10
50–59	4

(3)

Hours Worked	Pay
20	$160
25	$200
30	$240
35	$280

(2)

People	Number of Fingers
2	20
3	30
4	40
5	50

(4)

3. Which table does *not* show bivariate data?

Height (inches)	Weight (pounds)
39	50
48	70
60	90

(1)

Quiz Average	Frequency
70	12
80	15
90	6

(3)

Gallons	Miles Driven
15	300
20	400
25	500

(2)

Speed (mph)	Distance (miles)
40	80
50	120
55	150

(4)

4. Which situation is an example of bivariate data?
    (1) the number of pizzas Tanya eats during her years in high school
    (2) the number of times Ezra puts air in his bicycle tires during the summer
    (3) the number of home runs Elias hits per game and the number of hours he practices baseball
    (4) the number of hours Nellie studies for her mathematics tests during the first half of the school year

IX.   UNIVARIATE DATA

Dot Plots and Distributions

Key Terms and Concepts

A **dot plot** may be used to provide a graphic representation of a set of data. In a dot plot, each data value is represented by a dot, or by a similar symbol such as a plus sign (+) or asterisk (*). The dots are stacked on top of a number line.

For example: 29 students were asked to time their trip to school one morning and to round their results to the nearest multiple of 5 minutes. Their responses were plotted below. From the dot plot, we can see that two students took about 5 minutes each, one student took about 10 minutes, and so on.

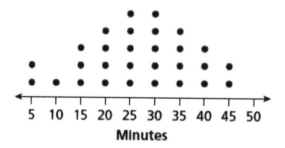

Dot plots tend to be used only for small data sets; for larger sets of data, histograms, which we'll see in the next section, are usually more efficient.

The **shape** of a dot plot can show the type of **distribution** of the data. If the data values are somewhat equally distributed, we call it a **uniform distribution**. In a **symmetrical distribution**, one could draw a vertical line on the dot plot that would divide it into two parts that are approximate mirror images of each other. A **skewed distribution** is neither uniform nor symmetric; the data is stacked mostly on the low end or on the high end of the dot plot.

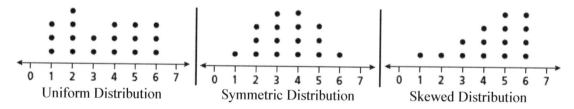

Uniform Distribution   Symmetric Distribution   Skewed Distribution

When the data tends to gather around a central value in a *symmetrical* dot plot, it is said to have a **normal distribution**, and the shape is often called a *bell curve*. The middle dot plot above is an example of a normal distribution of data; notice the bell-shaped curve as shown below:

In a **skewed distribution**, if the graph appears to have a tail to the right (and more of the data to the left), it is **skewed to the right**, or *positively skewed*. If it appears to have its data stretched to the left like a tail, it is considered **skewed to the left**, or *negatively skewed*.
For example: The following dot plot is skewed to the left.

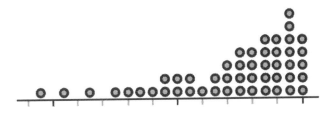

Model Problem
The following dot plot shows the fuel economy for a number of cars in miles per gallon (mpg).
How many cars have a fuel economy between 30 and 40 mpg inclusive?

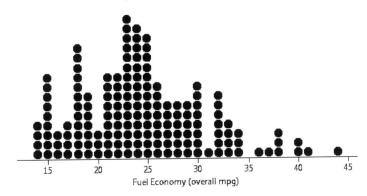

Fuel Economy (overall mpg)

Solution:

      30 cars

Explanation:

      Count the dots stacked vertically. *[There are 8 dots at the 30 mpg marker, 2 dots at the 40 mpg marker, and 20 more dots stacked between these two markers.]*

Practice Problems

1. Identify the distribution of the data represented by the dot plot below.	2. Draw a dot plot of data that has a symmetrical but *not* a normal distribution.
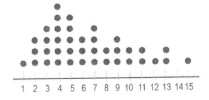	

Frequency Tables and Histograms

Key Terms and Concepts

Frequency is the number of times that a particular result or value occurs in a list of data. It is often helpful to count the frequencies using **tally** marks, with every 5 tally marks written as 卌 . Sometimes, data values are organized into **intervals** of equal sizes.

In a dot plot, the height of a stacked column of dots represents the frequency for that data value.

A **frequency histogram** is a type of bar graph where the height of each bar represents the number of data items in that interval. To create a frequency histogram:
1. First create a **frequency table** by tallying each item of data in the Tally column, then counting the tallies for each interval and writing the count in the Frequency column.
2. Create the frequency histogram by first labeling each interval along the horizontal axis and the range of frequencies along the vertical axis. The interval labels and the frequency labels must be equally spaced.
3. For each interval, draw a shaded rectangular bar as high as its corresponding frequency. There should be no spaces between the bars.

A **cumulative frequency histogram** is also a bar graph, but each bar represents the sum of all the frequencies from the *first* interval up to and including that interval. To create one:
1. Create a **cumulative frequency table** based on the frequency table. Each interval should be relabeled as spanning from the start of the *first* frequency table interval to the end of the current interval. The cumulative frequencies are the sums of the frequencies from the *first* interval up to and including that interval. *The last cumulative frequency should equal the number of data items.*
2. Create the cumulative frequency histogram by first labeling each interval along the horizontal axis and the range of cumulative frequencies along the vertical axis. The interval labels and the frequency labels must be equally spaced.
3. For each interval, draw a shaded rectangular bar as high as its corresponding cumulative frequency. There should be no spaces between the bars. *Each bar should be at least as tall as the bar to its left.*

To enter a set of data into the calculator:

1. Press [STAT] [Edit...] [ENTER].
2. If any values already appear in the [L1] column, select the column heading and press [CLEAR] [ENTER].
3. Enter the data values into the [L1] column.

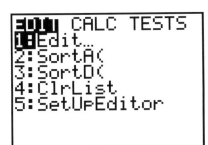

Note: If any columns other than [L1] appear in the table, you can delete them by selecting the column heading and pressing [DEL]. Also, if the [L1] column is missing, you can add it by selecting an empty column heading and then pressing [2nd] [L1] [ENTER].

To view a frequency histogram:

1. Press [2nd] [STAT PLOT] [ENTER].
2. Select [On] and press [ENTER].
3. Select the third Type of graph (histogram) and press [ENTER].
4. Press [GRAPH]. If you don't see the histogram, press [ZOOM] [9] (ZoomStat).

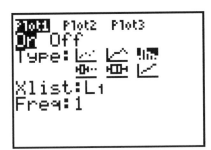

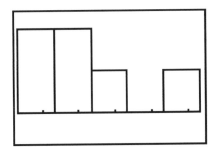

Model Problem

Create a frequency histogram and cumulative frequency histogram for the following set of 36 test scores, using intervals of 10 points.

14, 17, 28, 28, 30, 36, 45, 52, 58, 58, 61, 64, 64, 68, 68, 77, 77,
81, 81, 81, 81, 87, 87, 94, 94, 95, 95, 95, 95, 95, 95, 97, 97, 97, 100, 100

Solution:

(A) Interval	(B) Tally	(C) Frequency
0-10		0
11-20	‖	2
21-30	‖‖	3
31-40			1
41-50			1
51-60	‖‖	3
61-70	‖‖‖‖	5
71-80	‖	2
81-90	‖‖‖‖		6
91-100	‖‖‖‖ ‖‖‖‖ ‖‖‖	13

(D)

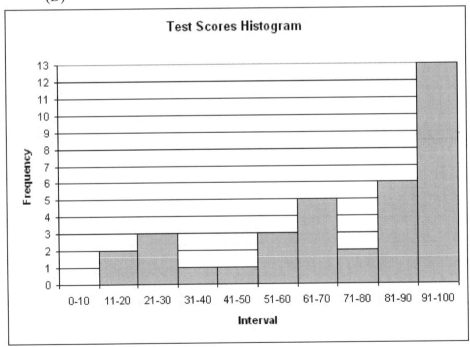

(Continued on next page...)

(Continued from previous page)

(E) Interval	(F) Cumulative Frequency
0-10	0
0-20	2
0-30	5
0-40	6
0-50	7
0-60	10
0-70	15
0-80	17
0-90	23
0-100	36

(G)

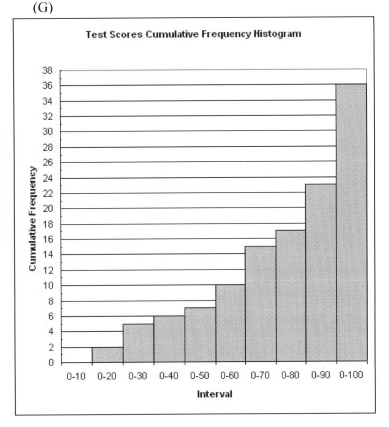

Explanation of Steps:
   (A) Create the frequency table using equally wide intervals. *[Each is 10 points wide.]*
   (B) Tally each data item by writing a tally mark for the corresponding interval.
   (C) Count the tally marks and write each count as a number in the Frequency column.
   (D) Draw the frequency histogram.
   (E) Create the cumulative frequency table based on the frequency table. Re-label the
       intervals so that they all start with the *first* interval's minimum value *[0]*.
   (F) Enter the cumulative frequencies. The first interval's cumulative frequency is equal to
       the first interval's frequency. The cumulative frequency for each interval after the first is
       the sum of the preceding cumulative frequency plus the corresponding interval's
       frequency. *[For example, the cumulative frequency for 0-30 is the sum of the cumulative
       frequency for 0-20, which is 2, plus the frequency of the 21-30 interval, which is 3.]*
   (G) Draw the cumulative frequency histogram.

Practice Problems

1. The accompanying histogram shows the heights of the students in Kyra's health class. What is the total number of students in the class?

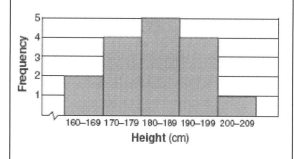

2. Casey talked to everyone in his apartment building to find out how many hours of television each person watched each day. The results are shown in the histogram below. Using the histogram, determine the total number of people in Casey's building.

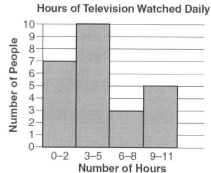

3. In the time trials for the 400-meter run at the state sectionals, the 15 runners recorded the times shown in the table below. Using the data from the frequency column, draw a frequency histogram on a separate grid.

400-Meter Run	
Time (sec)	Frequency
50.0–50.9	
51.0–51.9	II
52.0–52.9	JHT I
53.0–53.9	III
54.0–54.9	IIII

4. The following set of data represents the scores on a mathematics quiz:
58, 79, 81, 99, 68, 92, 76, 84, 53, 57, 81, 91, 77, 50, 65, 57, 51, 72, 84, 89
Complete the frequency table below and, on a separate grid, draw and label a frequency histogram of these scores.

Mathematics Quiz Scores

Interval	Tally	Frequency
50–59		
60–69		
70–79		
80–89		
90–99		

REGENTS QUESTIONS

Multiple Choice

1. The table below shows a cumulative frequency distribution of runners' ages.

   According to the table, how many runners are in their forties?
   (1) 25
   (2) 10
   (3) 7
   (4) 6

Cumulative Frequency Distribution of Runners' Ages

Age Group	Total
20–29	8
20–39	18
20–49	25
20–59	31
20–69	35

Constructed Response

2. Twenty students were surveyed about the number of days they played outside in one week. The results of this survey are shown below.

$$\{6,5,4,5,0,7,1,5,4,4,3,2,2,3,2,4,3,4,0,7\}$$

Complete the frequency table below for these data.

Number of Days Outside

Interval	Tally	Frequency
0–1		
2–3		
4–5		
6–7		

Complete the cumulative frequency table below using these data.

Number of Days Outside

Interval	Cumulative Frequency
0–1	
0–3	
0–5	
0–7	

On the grid below, create a cumulative frequency histogram based on the table you made.

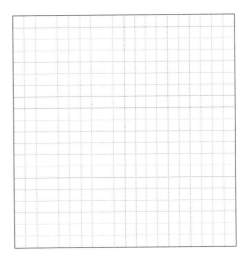

3. The Fahrenheit temperature readings on 30 April mornings in Stormville, New York, are shown below.

41°, 58°, 61°, 54°, 49°, 46°, 52°, 58°, 67°, 43°, 47°, 60°, 52°, 58°, 48°,
44°, 59°, 66°, 62°, 55°, 44°, 49°, 62°, 61°, 59°, 54°, 57°, 58°, 63°, 60°

Using the data, complete the frequency table below.

Interval	Tally	Frequency
40–44		
45–49		
50–54		
55–59		
60–64		
65–69		

On the grid below, construct and label a frequency histogram based on the table.

4. The diagram below shows a cumulative frequency histogram of the students' test scores in Ms. Wedow's algebra class.

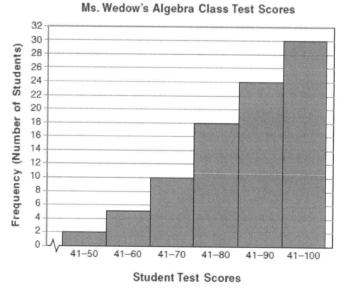

Determine the total number of students in the class. Determine how many students scored higher than 70. State which *ten-point interval* contains the median. State which *two ten-point* intervals contain the same frequency.

5. The test scores for 18 students in Ms. Mosher's class are listed below:

    86, 81, 79, 71, 58, 87, 52, 71, 87,
    87, 93, 64, 94, 81, 76, 98, 94, 68

Complete the frequency table below.

Draw and label a frequency histogram on the grid below.

Interval	Tally	Frequency
51–60		
61–70		
71–80		
81–90		
91–100		

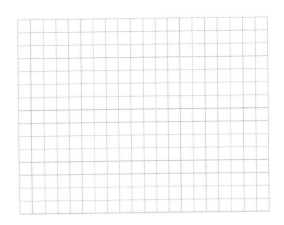

6. Ms. Hopkins recorded her students' final exam scores in the frequency table below.

Interval	Tally	Frequency
61–70	⊬⊬	5
71–80	IIII	4
81–90	⊬⊬ IIII	9
91–100	⊬⊬ I	6

On the grid below, construct a frequency histogram based on the table.

7. The following cumulative frequency histogram shows the distances swimmers completed in a recent swim test.

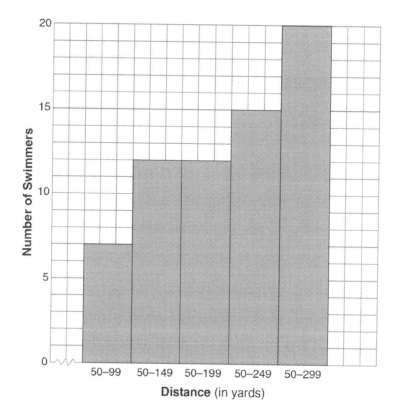

Based on the cumulative frequency histogram, determine the number of swimmers who swam between 200 and 249 yards. Determine the number of swimmers who swam between 150 and 199 yards. Determine the number of swimmers who took the swim test.

8. The cumulative frequency table below shows the number of minutes 31 students spent text messaging on a weekend.

Text-Use Interval (minutes)	Cumulative Frequency
41–50	2
41–60	5
41–70	10
41–80	19
41–90	31

Determine which 10-minute interval contains the median. Justify your choice.

Central Tendency

Key Terms and Concepts

For the definitions below, we will refer to this sample set of quantitative data:
   2, 4, 8, 3, 8, 8, 2

The lowest value in the data set is called the **minimum**, the highest value is called the **maximum**, and the **range** is the *difference* between the *maximum* and *minimum* values. Range is actually a measure of **spread**, not central tendency, in that it measures how spread out the data set is.
For example: For the sample data, 2 is the minimum, 8 is the maximum, and the range is 6.

The three most commonly used measures of **central tendency** are the mean, median and mode.

The **mean** (often written as $\bar{x}$) is calculated as a numerical average, by *adding* all the numbers in the data and *dividing* this sum by how many data values there are.
For example: The mean of the sample set of data is $\dfrac{2+4+8+3+8+8+2}{7} = \dfrac{35}{7} = 5$

To find the **median** for a set of data, we must first arrange the data in ascending order.
For an *odd* number of data values, the median is the *middle number*.
For example: The sample set of data is arranged as: 2, 2, 3, 4, 8, 8, 8.
                There are 7 data values. The median, or middle number, is 4.
For an *even* number of values, the median is the *mean* (average) of the *two middle numbers*.
For example: Suppose we add another 2 to the data: 2, 2, 2, 3, 4, 8, 8, 8.
                Now there are 8 values. The median is the average of 3 and 4, or 3.5.

The **mode** is the data value that appears *most often*. It is helpful to arrange the data in ascending order before determining the mode.
For example: The sample set of data is arranged as: 2, 2, 3, 4, 8, 8, 8.
                Since the value 8 appears most often (three times), the mode is 8.
It is possible to have more than one mode. A set of data with two modes is called **bimodal**.
For example: Suppose we add another 2 to the sample data: 2, 2, 2, 3, 4, 8, 8, 8.
                Now there are two modes, 2 and 8, since both numbers appear three times.
If each number appears with the same frequency, then there is **no mode**.
For example: This set of data has no mode: 2, 2, 3, 3, 4, 4.

Sometime, one of the measures of central tendency is a better indicator than another. In some cases there are one or a few data values, called **outliers**, which are much higher or lower than the rest of the data. Outliers can skew the mean.
For examples: Consider these two sets of data.
                2, 3, 5, 5, 7, 8, 33   2, 3, 5, 5, 25, 25, 33
                Mean = 9   Mean = 14
                Median = 5   Median = 5
                In the first set of data, an outlier (33) skews the mean so that the mean (9) is actually larger than 6 of the 7 data values. In this case, the median may be a better indicator of central tendency. However, in the second set of data, the mean (14) may be more appropriate, since nearly half the values are at least 25.

181

A *skewed distribution* can easily be seen in a frequency histogram. The histogram on the left shows a somewhat *symmetrical distribution*, in which the left and right sides have approximately the same shape. The histogram on the right is skewed to the right (*positively skewed*), in that it has a "tail" to the right and more of the data to the left.

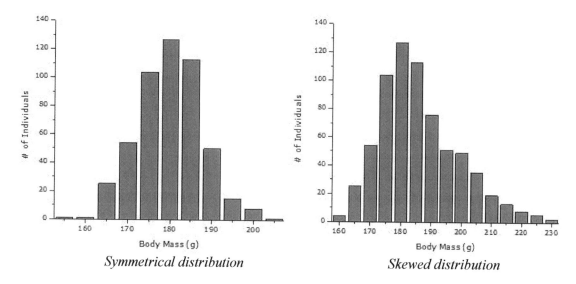

Symmetrical distribution   *Skewed distribution*

When the data has a *symmetrical* distribution, the *mean* is usually a better measure of central tendency, but when the data is *skewed*, the *median* is usually a better indicator. For a more informative analysis, the *standard deviation* can be calculated with the mean, or the *quartiles* may be calculated with the median; these will be presented in later sections.

If we **add the same constant to each value** in a set of data, this will add that constant to the mean, median, and mode. The range will not change.

For example: Test scores in a class are originally 55, 60, 62, 70, 70.
The mean = 63.4, median = 62, and mode = 70.
The teacher decides to "curve" the test by adding 10 points to each score.
The data set becomes 65, 70, 72, 80, 80. Each new measure of central tendency
is now 10 points higher: mean = 73.4, median = 72, and mode = 80.

The same is true if we **multiply each data value by the same constant** (other than zero). The mean, median and mode will also be multiplied by the same constant factor as a result. However, in this case the range will also multiplied by the same factor.

To find the measures of central tendency on the calculator:

1. First, enter the data into the L1 column of the $\boxed{\text{STAT}}$ screen (see page 173).
2. Press $\boxed{\text{STAT}}$, select $\boxed{\text{CALC}}$ and press $\boxed{1}$ (1-Var Stats) and $\boxed{\text{ENTER}}$.
3. The mean is the value of $\boxed{\overline{\text{X}}}$. Press the down arrow to $\boxed{\text{Med}}$ for the median. Unfortunately, the calculator does not calculate the mode.

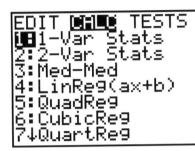

```
EDIT CALC TESTS
1:1-Var Stats
2:2-Var Stats
3:Med-Med
4:LinReg(ax+b)
5:QuadReg
6:CubicReg
7↓QuartReg
```

```
1-Var Stats L1■
```

```
1-Var Stats
 x̄=27.33333333
 Σx=164
 Σx²=5282
 Sx=12.64383908
 σx=11.54219313
↓n=6
```

```
1-Var Stats
↑n=6
 minX=12
 Q₁=18
 Med=26
 Q₃=34
 maxX=48
```

An alternate method of finding the mean (or median) is to press $\boxed{\text{2nd}}$ $\boxed{\text{STAT}}$, then select $\boxed{\text{MATH}}$. Press $\boxed{3}$ for mean (or $\boxed{4}$ for median), press $\boxed{\text{2nd}}$ $\boxed{1}$ for L1, then $\boxed{)}$ and $\boxed{\text{ENTER}}$.

Model Problem 1: *calculating measures of central tendency*
For the given set of data, find the mean, median, and mode: 1, 3, 5, 7, 9, 13, 21, 25, 25, 31.

Solution:

(A) Mean $= \dfrac{1+3+5+7+9+13+21+25+25+31}{10} = \dfrac{140}{10} = 14$

(B) Median $= \dfrac{9+13}{2} = 11$

(C) Mode $= 25$

Explanation of Steps:

(A) Find the mean by adding all the values and dividing this sum *[140]* by the number of data values *[10]*.

(B) For an even number of data values *[there are 10 values in this set]*, we need to take the mean (numerical average) of the middle two numbers *[9 and 13]*.

(C) The mode is the value that appears most often. *[Only 25 appears more than once.]*

Model Problem 2: *distribution and shape*

The dotplot below shows the number of televisions owned by each family on a city block.

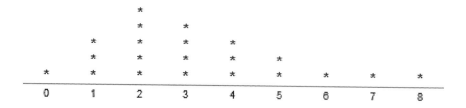

Which of the following statements are true?
  (1) The distribution is right-skewed with no outliers.
  (2) The distribution is right-skewed with one outlier.
  (3) The distribution is left-skewed with no outliers.
  (4) The distribution is left-skewed with one outlier.
  (5) The distribution is symmetric.

Solution:
The correct answer is (1).

Explanation:
Most of the observations are on the left side of the distribution, so the distribution is right-skewed. And none of the observations is extreme, so there are no outliers.

Practice Problems

1. For a school report, Luke contacted a car dealership to collect data on recent sales. He asked, "What color do buyers choose most often for their car?" White was the response. What statistical measure does the response "white" represent?	2. The weights of all the students in grade 9 are arranged from least to greatest. Which statistical measure separates the top half of this set of data from the bottom half?
3. Which of the following sets of data is bimodal? a) 1, 1, 2, 5, 5, 6 b) 1, 1, 2, 2, 2, 3 c) 1, 2, 3, 4, 5, 6 d) 1, 1, 2, 2, 3, 3	4. Each value in a set of data is divided by two. How does this affect the mean, median and mode for this set of data?
5. The table shows the high and low temperatures for five cities. Which city had the greatest temperature range? TEMPERATURES ON OCTOBER 1ST FOR FIVE CITIES (in °F) 		High	Low	 |---|---|---| 	City A	72	50	 	City B	90	75	 	City C	83	72	 	City D	50	37	 	City E	92	72		6. Sara's test scores in mathematics were 64, 80, 88, 78, 60, 92, 84, 76, 86, 78, 72, and 90. Determine the mean, the median, and the mode of Sara's test scores.
7. The accompanying graph shows the high temperatures in Elmira, New York, for a 5-day period in January. Find the mean, median, and mode. 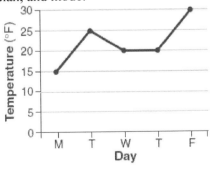	8. The table below shows the distribution of bowling scores. In which interval does the median lie? 	Interval	Frequency	 	---	---	 	91–110	10	 	111–130	11	 	131–150	8	 	151–170	4	 	171–190	6	 	191–210	5	

REGENTS QUESTIONS

Multiple Choice

1. The set of numbers {4,7,12} has

    (1) a range of 3 and a median of 7   (3) a range of 12 and a median of $7\frac{2}{3}$

    (2) a range of 8 and a median of 7   (4) a range of 8 and a median of $7\frac{2}{3}$

2. Melissa's test scores are 75, 83, and 75. Which statement is true about this set of data?
    - (1) mean < mode
    - (2) mode < median
    - (3) mode = median
    - (4) mean = median

3. Which statement is true about the data set 3, 4, 5, 6, 7, 7, 10?
    - (1) mean = mode
    - (2) mean > mode
    - (3) mean = median
    - (4) mean < median

4. Alex earned scores of 60, 74, 82, 87, 87, and 94 on his first six algebra tests. What is the relationship between the measures of central tendency of these scores?
    - (1) median < mode < mean
    - (2) mean < mode < median
    - (3) mode < median < mean
    - (4) mean < median < mode

5. This year, John played in 10 baseball games. In these games he had hit the ball 2, 3, 0, 1, 3, 2, 4, 0, 2, and 3 times. In the first 10 games he plays next year, John wants to increase his average (mean) hits per game by 0.5. What is the total number of hits John needs over the first 10 games next year to achieve his goal?
    - (1) 5
    - (2) 2
    - (3) 20
    - (4) 25

6. Sam's grades on eleven chemistry tests were 90, 85, 76, 63, 94, 89, 81, 76, 78, 69, and 97. Which statement is true about the measures of central tendency?
    - (1) mean > mode
    - (2) mean < median
    - (3) mode > median
    - (4) median = mean

7. Which statement is true about the data set 4, 5, 6, 6, 7, 9, 12?
    - (1) mean = mode
    - (2) mode = median
    - (3) mean < median
    - (4) mode > mean

8. Mr. Taylor raised all his students' scores on a recent test by five points. How were the mean and the range of the scores affected?
    - (1) The mean increased by five and the range increased by five.
    - (2) The mean increased by five and the range remained the same.
    - (3) The mean remained the same and the range increased by five.
    - (4) The mean remained the same and the range remained the same.

Constructed Response

9. The test scores for five students were 59, 60, 63, 76, and 87. How many points greater than the median is the mean?

10. Two social studies classes took the same current events examination that was scored on the basis of 100 points. Mr. Wong's class had a median score of 78 and a range of 4 points, while Ms. Rizzo's class had a median score of 78 and a range of 22 points. Explain how these classes could have the same median score while having very different ranges.

11. The values of 11 houses on Washington St. are shown in the table below.

Value per House	Number of Houses
$100,000	1
$175,000	5
$200,000	4
$700,000	1

Find the mean value of these houses in dollars. Find the median value of these houses in dollars. State which measure of central tendency, the mean or the median, *best* represents the values of these 11 houses. Justify your answer.

12. The prices of seven race cars sold last week are listed in the table below.

Price per Race Car	Number of Race Cars
$126,000	1
$140,000	2
$180,000	1
$400,000	2
$819,000	1

What is the mean value of these race cars, in dollars? What is the median value of these race cars, in dollars? State which of these measures of central tendency best represents the value of the seven race cars. Justify your answer.

13. Ms. Mosher recorded the math test scores of six students in the table below.

Student	Student Score
Andrew	72
John	80
George	85
Amber	93
Betty	78
Roberto	80

Determine the mean of the student scores, to the *nearest tenth*. Determine the median of the student scores. Describe the effect on the mean and the median if Ms. Mosher adds 5 bonus points to each of the six students' scores.

14. Given the following list of students' scores on a quiz:

5, 12, 7, 15, 20, 14, 7

Determine the median of these scores. Determine the mode of these scores. The teacher decides to adjust these scores by adding three points to each score. Explain the effect, if any, that this will have on the median and mode of these scores.

Standard Deviation

Key Terms and Concepts

The **standard deviation (SD)** is the most commonly used measure of the **spread** (also known as *variability* or *dispersion*) of a set of data. A *low* standard deviation means the data values tend to be *close to the mean*, while a *high* standard deviation means the the values are *more spread out*. In other words, more **consistent** data values should result in a **lower** standard deviation.

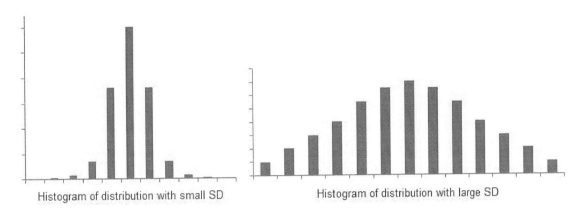

Histogram of distribution with small SD   Histogram of distribution with large SD

As discussed earlier, in the section called *Population and Sample*, our data usually represents a *sample* rather than the entire *population*. The methods for calculating the SD for a sample and for a population are slightly different. We will usually calculate the **sample standard deviation**, which is an estimate of the *population standard deviation*.

To calculate the sample standard deviation of *n* data values:
1) Calculate the mean of the *n* values.
2) For each data value, find the square of the difference (*deviation*) from the mean.
3) Find the *sample variance*, which is the sum of these squares divided by $n-1$.
4) Take the square root of the sample variance.

This can be expressed as a formula: $s = \sqrt{\dfrac{\sum (x-\bar{x})^2}{n-1}}$ (where $\sum$ means the *sum of each*)

To find the sample standard deviation on the calculator:

1. First, enter the data into the L1 column of the $\boxed{\text{STAT}}$ screen (see page 173).
2. Press $\boxed{\text{STAT}}$, select $\boxed{\text{CALC}}$ and press $\boxed{1}$ (1-Var Stats) and $\boxed{\text{ENTER}}$.
3. The standard deviation is the value of $\boxed{\text{Sx}}$.

```
EDIT CALC TESTS
1▮1-Var Stats
2:2-Var Stats
3:Med-Med
4:LinReg(ax+b)
5:QuadReg
6:CubicReg
7↓QuartReg
```

```
1-Var Stats
 x̄=27.33333333
 Σx=164
 Σx²=5282
 Sx=12.64383908
 σx=11.54219313
↓n=6
```

When the data represents the entire population, which is rarely the case, you should calculate the **population standard deviation** instead. The population SD is often symbolized by the Greek letter σ (sigma). It is calculated the same way as the sample SD except that the *population variance* is calculated by dividing the sum of the squares by n, not by $n-1$. (It is the mean of the squares.) On the calculator, it is designated by $\boxed{\sigma x}$.

The reason we divide by $n-1$ for the sample variance, rather than simply dividing by n as we do for the population variance, is to compensate for the fact that *sample* data will usually be closer to its own mean than to the actual (but unknown) mean of the entire *population*. This is known as Bessel's correction, named after Friedrich Bessel, a German astronomer and mathematician.

Model Problem 1: *conceptualizing standard deviation*
In the following chart, stock prices for companies UCX and SWC are recorded at the start of each month over a seven month period. Both stocks have the same mean price of $70 over that time. Which stock's monthly prices have a greater standard deviation, according to the graph?

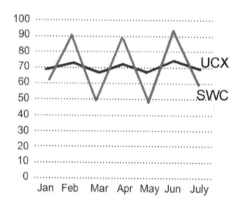

Solution:
SWC stock has a larger standard deviation.

Explanation:
The data set with the greater variability will have the higher standard deviation. *[The UCX prices are much more consistent, with prices closer to the mean, whereas the SWC prices fluctuate greatly, with monthly prices that are more distant from the mean.]*

Model Problem 2: *calculating standard deviation*

Find the sample standard deviation of this set of data.

2, 4, 4, 4, 5, 5, 5, 7, 9

Solution:

(A) $\dfrac{2+4+4+4+5+5+5+7+9}{9} = \dfrac{45}{9} = 5$

(B)

value	(value – mean)^2	result
2	$(2-5)^2$	9
4	$(4-5)^2$	1
4	$(4-5)^2$	1
4	$(4-5)^2$	1
5	$(5-5)^2$	0
5	$(5-5)^2$	0
5	$(5-5)^2$	0
7	$(7-5)^2$	4
9	$(9-5)^2$	16

(C) $\dfrac{9+1+1+1+0+0+0+4+16}{9-1} = \dfrac{32}{8} = 4$

(D) The standard deviation $= \sqrt{4} = 2$.

Explanation of steps:

(A) Calculate the mean. *[5]*.
(B) For each data value, find the square of the difference from the mean.
(C) Find the variance by dividing the sum of the squares by $n-1$ *[32 ÷ 8]*.
(D) Take the square root of the variance *[2]*.

Practice Problems

1. Find the mean and sample standard deviation to the *nearest tenth*. 22, 99, 102, 33, 57, 75, 100, 81, 62, 29	2. Find the mean and sample standard deviation to the *nearest tenth*. 35, 50, 60, 60, 75, 65, 80
3. Find the mean and sample standard deviation to the *nearest tenth*. 51, 48, 47, 46, 45, 43, 41, 40, 40, 39	4. Find the mean and sample standard deviation to the *nearest tenth*. 11, 7, 14, 2, 8, 13, 3, 6, 10, 3, 8, 4, 8, 4, 7
5. This chart shows the weekly salary of five employees working at company ABC. Employee Number | Salary 3201 | $612 2734 | $588 2461 | $604 3582 | $625 3144 | $621 Find the mean and sample standard deviation of this data.	6. The mean for a set of data is 8.9 and the standard deviation is 1. The mean for a second set of data is 8.9 and the standard deviation is 2. In which data set do the values cluster closer to the mean?

For problem 5, the table is:

Employee Number	Salary
3201	$612
2734	$588
2461	$604
3582	$625
3144	$621

REGENTS QUESTIONS

Multiple Choice

1. Tanner and Robbie discovered that the means of their grades for the first semester in Mrs. Merrell's mathematics class are identical. They also noticed that the standard deviation of Tanner's scores is 20.7, while the standard deviation of Robbie's scores is 2.7. Which statement must be true?
    (1) In general, Robbie's grades are lower than Tanner's grades.
    (2) Robbie's grades are more consistent than Tanner's grades.
    (3) Robbie had more failing grades during the semester than Tanner had.
    (4) The median for Robbie's grades is lower than the median for Tanner's grades.

2. Jean's scores on five mathematics tests were 98, 97, 99, 98, and 96. Her scores on five English tests were 78, 84, 95, 72, and 79. Which statement is true about the standard deviations for the scores?
    (1) The standard deviation for the English scores is greater than the standard deviation for the math scores.
    (2) The standard deviation for the math scores is greater than the standard deviation for the English scores.
    (3) The standard deviations for both sets of scores are equal.
    (4) More information is needed to determine the relationship between the standard deviations.

Constructed Response

3. On a nationwide examination, the Adams School had a mean score of 875 and a standard deviation of 12. The Boswell School had a mean score of 855 and a standard deviation of 20. In which school was there greater consistency in the scores? Explain how you arrived at your answer.

4. Find the standard deviation, to the *nearest hundredth*, for the following measurements:
    24, 28, 29, 30, 30, 31, 32, 32, 32, 33, 35, 36

5. The scores on a mathematics test are:
    42, 51, 58, 64, 70, 76, 76, 82, 84, 88, 88, 90, 94, 94, 94, 97
    For this set of data, find the standard deviation to the *nearest tenth*.

6. During a 10-game season, a high school football team scored the following number of points: 14, 17, 21, 10, 35, 27, 13, 7, 45, 21. Find the standard deviation of these scores to the *nearest thousandth*.

7. For these measurements, find the standard deviation, to the *nearest hundredth*:
    85, 88, 79, 79, 80, 92, 94, 78, 80, 85

Percentiles

Key Terms and Concepts

A value's **percentile** is the percent of values in a set of data that lie at or below the given value. The percent is generally rounded to the nearest whole and the given data value is stated as being "at the pth percentile" or as having "a percentile rank of p". The lowest value is called the first (*not* the zero) percentile.

For example: Andy scores an 85 on an exam, which is better than 300 of the 325 students who took the exam. Andy's score is the 92nd percentile. ($300 \div 325 \approx 92.3\%$.)

If given a set of data, find a given value's percentile p by the following steps:
1. Arrange the data in ascending order.
2. Count how many values, n, are in the set.
3. Count how many values, b, are below the given value.
4. Count how many times the given value appears in the data and divide this number by 2. We'll call this the half frequency, h.
5. Calculate $p = \dfrac{b+h}{n}$ and round the result to the nearest percent.

For example: To find the percentile rank of 10 in the given set of data:

2, 4, 8, 10, 14, 18, 20, 22, 30, 32, 38

There are 11 values, so $n = 11$. There are 3 values below 10, so $b = 3$.

There is only one 10, so $h = \dfrac{1}{2}$ or 0.5.

$$p = \frac{b+h}{n} = \frac{3+0.5}{11} = \frac{3.5}{11} = 0.3\overline{18} \approx 32\% \text{ or the } 32^{nd} \text{ percentile.}$$

Model Problem 1: *data values not given*

If Adrienne graduated with the 40th highest GPA out of a class of 300 students, what was her percentile rank?

Solution:

$\dfrac{260}{300} = 0.8\overline{6} \approx 87\%$. Adrienne's GPA is at the 87th percentile.

Explanation of Steps:

Find the number of values below the given value *[300 – 40 = 260]*, and divide by the total number of values in the set of data to find the percent. *[Since 260 students' GPAs fell below hers, $260 \div 300 \approx 86.7\%$.]* Round to the nearest whole percentile.

Practice Problems

1. Dawn scored higher than 22 out of the 30 students in her class. What was her percentile rank?	2. Tony's GPA is at the 40th percentile among the 600 students in his senior class. How many seniors have GPAs above Tony's?

Model Problem 2: *data values given*

Find the percentile rank of 65 in the following set of test scores:

   70, 65, 80, 58, 92, 85, 65, 50, 80, 58

Solution:

(A) 50, 58, 58, 65, 65, 70, 80, 80, 85, 92

(B) $n = 10$ (C) $b = 3$ (D) $h = 1$

(E) $p = \dfrac{b+h}{n} = \dfrac{3+1}{10} = \dfrac{4}{10} = 40\%$. The value 65 is at the 40th percentile.

Explanation of Steps:

(A) Arrange the data in ascending order.

(B) Count how many values, n, are in the set.

(C) Count how many values, b, are below the given value.

(D) Count how many times the given value appears in the data and divide this number by 2. We'll call this the half frequency, h. [*There are two 65's, so h = 2/2 = 1.*]

(E) Calculate $p = \dfrac{b+h}{n}$ and round the result to the nearest percent.

Practice Problems

3. For the given set of data, find the percentile rank of the value 70. 25, 90, 87, 58, 42, 95, 120, 64, 75, 39, 70	4. From the given frequency table of test scores, find the percentile rank of 90.

Score	Frequency
60	1
70	9
80	8
90	2
100	5

REGENTS QUESTIONS

Multiple Choice

1. Brianna's score on a national math assessment exceeded the scores of 95,000 of the
    125,000 students who took the assessment. What was her percentile rank?

    (1) 6   (3) 31
    (2) 24   (4) 76

Quartiles

Key Terms and Concepts

For a set of data, the value at the 25th percentile is called the **first quartile** or **lower quartile** (and is often denoted as Q_1). The value at the 50th percentile is called the **second quartile** or **median** (or Q_2), and the value at the 75th percentile is called the **third quartile** or **upper quartile** (or Q_3).

The difference between the upper and lower quartiles is called the **interquartile range** (or IQR). The interquartile range is another measure of spread, along with range and standard deviation.

The IQR can be used to identify *outliers*, which are extremely high or low values in the data. A data value should be considered an outlier if it is more than $1.5 \times$ IQR below the lower quartile or above the upper quartile.

To identify the quartiles for a set of data:
1. Arrange the data in ascending order.
2. First find the *median (or second quartile)*, and label it Q_2. It is the middle value (for an odd number of data values), or the average of the two middle values (for an even number of data values).
    > **Use this tip:** For an *odd number* of data values, *circle* the middle value, but for an *even number* of data values, *draw a vertical line* between the two middle values. When we find the quartiles in the next steps, we will ignore a circled value but include values to the left or right of a line.
3. Find the *lower (first) quartile* by looking only at the subgroup of values that are to the *left* of the middle (that is, to the left of a circled value or line). The median of this subgroup is the lower quartile.
4. Find the *upper (third) quartile* by looking only at the subgroup of values that are to the *right* of the middle (to the right of a circled value or line). The median of this subgroup is the upper quartile.

To find the quartiles on the calculator:
1. First, enter the data into the L1 column of the $\boxed{\text{STAT}}$ screen (see page 173).
2. Press $\boxed{\text{STAT}}$ $\boxed{\text{CALC}}$ $\boxed{\text{1-Var Stats}}$. Press $\boxed{\text{2nd}}$ [L1] $\boxed{\text{ENTER}}$.
3. Scroll down to find the number of data values $\boxed{\text{n}}$, the minimum $\boxed{\text{minX}}$, first quartile $\boxed{\text{Q1}}$, median $\boxed{\text{Med}}$, third quartile $\boxed{\text{Q3}}$, and maximum $\boxed{\text{maxX}}$.

```
EDIT CALC TESTS
1▶1-Var Stats
2:2-Var Stats
3:Med-Med
4:LinReg(ax+b)
5:QuadReg
6:CubicReg
7↓QuartReg
```
```
1-Var Stats
↑n=6
 minX=12
 Q₁=18
 Med=26
 Q₃=34
 maxX=48
```

197

Model Problem

Identify the first, second and third quartiles and calculate the interquartile range for the following set of ten data values:

   86, 72, 85, 89, 86, 92, 73, 71, 91, 82

Solution:

 (A)

$$71, \quad 72, \quad \text{⑦③} \quad 82, \quad 85, \mid 86, \quad 86, \quad \text{⑧⑨} \quad 91, \quad 92$$

 (B) Q_2 (median) = 85.5
 (C) $Q_1 = 73$ and $Q_3 = 89$
 (D) IQR = $89 - 73 = 16$

Explanation of Steps:

 (A) Arrange the data in ascending order.
 (B) Find the median. *[Since there is an even number of values, we draw a line between the two middle values, 85 and 86, and calculate the median (or second quartile) as the average of these two middle values.]*
 (C) Find the lower and upper quartiles. *[For the subgroup of 5 values to the left of the line, the middle number (73) is the lower quartile. For the subgroup of 5 values to the right of the line, the middle number (89) is the upper quartile.]*
 (D) Calculate the interquartile range (IQR) as the difference between Q_3 and Q_1.

Practice Problems

1. Find the first, second and third quartiles for the following set of data: 5, 6, 7, 8, 12, 14, 17, 17, 18, 19, 19	2. Find the first, second and third quartiles for the following set of data: 3, 6, 7, 7, 8, 9, 9, 9, 10, 12, 13, 15
3. Find the first, second and third quartiles, and the interquartile range, for the following set of data: 33, 28, 45, 21, 32, 53, 41, 28, 50	4. For the set of test scores shown by the frequency table below, find the first, second and third quartiles. <table><tr><td>Score</td><td>Frequency</td></tr><tr><td>60</td><td>1</td></tr><tr><td>70</td><td>9</td></tr><tr><td>80</td><td>8</td></tr><tr><td>90</td><td>2</td></tr><tr><td>100</td><td>5</td></tr></table>

REGENTS QUESTIONS

Multiple Choice

1. The freshman class held a canned food drive for 12 weeks. The results are summarized in the table below.

Canned Food Drive Results

Week	1	2	3	4	5	6	7	8	9	10	11	12
Number of Cans	20	35	32	45	58	46	28	23	31	79	65	62

Which number represents the second quartile of the number of cans of food collected?
   (1) 29.5   (3) 40
   (2) 30.5   (4) 60

2. The cumulative frequency table below shows the length of time that 30 students spent text messaging on a weekend.

Minutes Used	Cumulative Frequency
31–40	2
31–50	5
31–60	10
31–70	19
31–80	30

Which 10-minute interval contains the first quartile?
   (1) 31 – 40   (3) 51 – 60
   (2) 41 – 50   (4) 61 – 70

3. Christopher looked at his quiz scores shown below for the first and second semester of his Algebra class.
   Semester 1: 78, 91, 88, 83, 94
   Semester 2: 91, 96, 80, 77, 88, 85, 92
Which statement about Christopher's performance is correct?
   (1) The interquartile range for semester 1 is greater than the interquartile range for semester 2.
   (2) The median score for semester 1 is greater than the median score for semester 2.
   (3) The mean score for semester 2 is greater than the mean score for semester 1.
   (4) The third quartile for semester 2 is greater than the third quartile for semester 1.

<u>*Constructed Response*</u>

4. The heights, in inches, of 10 high school varsity basketball players are 78, 79, 79, 72, 75, 71, 74, 74, 83, and 71. Find the interquartile range of this data set.

Box Plots

Key Terms and Concepts

A **box-and-whisker plot** (or **box plot**) is a used to represent a set of data graphically. The "box" marks the values of the *first, second,* and *third quartiles* of the data set. The "whiskers" show the *minimum* (lowest) and *maximum* (highest) values of the data set.

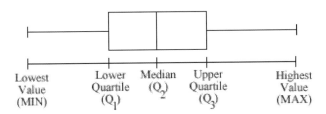

The lowest value in the data set is called the **minimum**, the highest value is called the **maximum**, and the **range** is the *difference* between the *maximum* and *minimum* values.

By the definition of quartiles, each section of a box plot represents 25% (one quarter) of the data values. 25% of the data is at or below the lower quartile, 50% is at or below the median, and 75% is at or below the upper quartile.

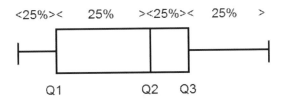

If the data has **symmetrical** (*normal*) distribution, the whiskers will be about the same length and the median will be at about the center of the data. For a **skewed** distribution, the whisker on the side of the "tail" will be longer and the median will be closer to the shorter whisker.

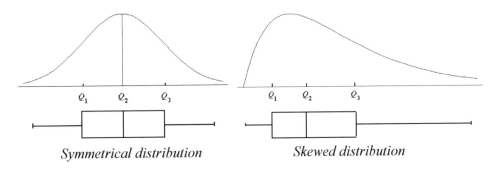

Symmetrical distribution   *Skewed distribution*

To create a box plot:
1. Arrange the data from smallest to largest.
2. Draw a number line from the minimum (or below) to the maximum (or above).
3. Find the median and plot it as a line.
4. Find the lower and upper quartiles, plot them as lines, and connect the box.
5. Plot the minimum and maximum as smaller lines and connect the whiskers.

The calculator can be used to display a box plot.

1. First, enter the data into the L1 column of the [STAT] screen (see page 173).

2. Press [2nd] [STAT PLOT] [Plot1...] [ENTER]. Select [On] [ENTER]. For the [Type], move right to select the fifth chart type ⊡ and check that the [Xlist] is [L1].

3. Press [ZOOM] [ZBox]. The plot will appear.

4. Press [TRACE] and then the left and right arrow keys to see the five summary numbers: minX, Q1, Med, Q2, and maxX.

Box plots with outliers: Sometimes, outliers are drawn separately as dots (or similar symbols such as asterisks) on a box plot and excluded from the lengths of the whiskers. Remember, an outlier is considered to be any data value that is more than $1.5 \times IQR$ above the upper quartile or more than $1.5 \times IQR$ below the lower quartile. *IQR* is the interquartile range.

For example: Consider the data set, {20, 25, 25, 27, 28, 31, 33, 34, 36, 37, 44, 50, 59, 85, 86}. The quartiles are 27, 34, and 50, so the box plot would look like this:

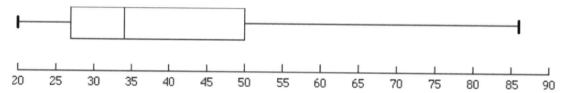

Notice the long right whisker due to the fact that two outliers, 85 and 86, are included. We know these are outliers because they are more than $1.5 \times IQR$ above the upper quartile.

$$IQR = Q_3 - Q_1 = 50 - 27 = 23$$

$$Q_3 + (1.5 \times IQR) = 50 + (1.5 \times 23) = 50 + 34.5 = 84.5$$

So, any values above 84.5 are outliers. Therefore, 85 and 86 are outliers.

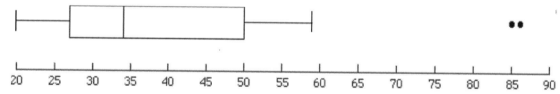

If we show these outliers as dots, the box plot would appear as above. The right whisker is shortened in length to the next largest (*adjacent*) value below the outliers, which happens to be 59. This now shows a big gap below the outliers.

Model Problem

Create a box plot for the following data:

2, 4, 8, 10, 14, 18, 20, 22, 30, 32, 38

Solution:

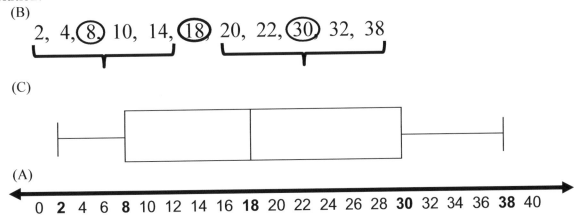

(B)

(C)

(A)

Explanation of Steps:

(A) Once the data is arranged in ascending order *[as this data already is]*, draw a number line from below the minimum *[2]* to above the maximum *[38]*, and label the intervals. *[Our number line can go from 0 to 40 in 2-unit intervals.]*

(B) Find the median *[18]* and the lower and upper quartiles *[8 and 30]*.

(C) Plot the median as a line *[over 18]* and the lower and upper quartiles as lines *[over 8 and 30]* and connect the box. Add whiskers (shorter lines) over the minimum *[2]* and maximum *[38]* and connect them to the box as whiskers.

Practice Problems

1. The accompanying diagram shows a box plot of student test scores. What is the median score?

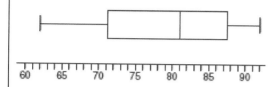

2. The accompanying box plot represents the scores earned on a science test. What is the median score?

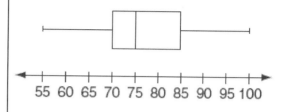

3. Create a box plot for the following data:

   89, 73, 84, 91, 87, 77, 94

4. Create a box plot for the following data:

   65, 75, 92, 84, 62, 96, 88, 79, 82

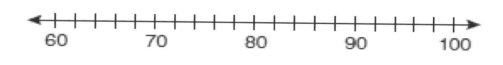

REGENTS QUESTIONS

Multiple Choice

1. The data set 5, 6, 7, 8, 9, 9, 9, 10, 12, 14, 17, 17, 18, 19, 19 represents the number of hours spent on the Internet in a week by students in a mathematics class. Which box-and-whisker plot represents the data?

(1)
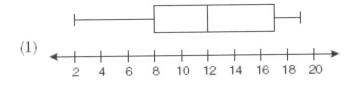

(2)
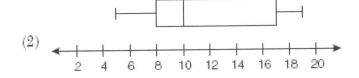

(3)
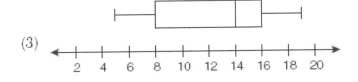

(4)
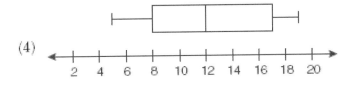

2. What is the value of the third quartile shown on the box-and-whisker plot below?

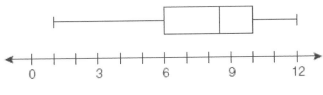

(1) 6   (3) 10
(2) 8.5   (4) 12

3. A movie theater recorded the number of tickets sold daily for a popular movie during the month of June. The box-and-whisker plot shown below represents the data for the number of tickets sold, in hundreds.

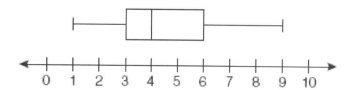

Which conclusion can be made using this plot?
   (1) The second quartile is 600.
   (2) The mean of the attendance is 400.
   (3) The range of the attendance is 300.
   (4) Twenty-five percent of the attendance is between 300 and 400.

4. The box-and-whisker plot below represents students' scores on a recent English test.

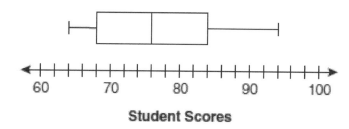

Student Scores

What is the value of the upper quartile?
   (1) 68   (3) 84
   (2) 76   (4) 94

5. The box-and-whisker plot below represents the math test scores of 20 students.

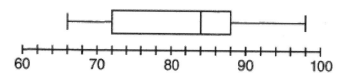

What percentage of the test scores are *less than* 72?
   (1) 25   (3) 75
   (2) 50   (4) 100

6. What is the range of the data represented in the box-and-whisker plot shown below?

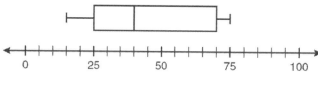

   (1) 40   (3) 60
   (2) 45   (4) 100

7. Based on the box-and-whisker plot below, which statement is *false*?

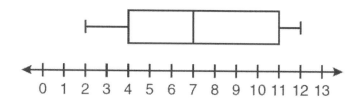

    (1) The median is 7.   (3) The first quartile is 4.
    (2) The range is 12.   (4) The third quartile is 11.

8. The box-and-whisker plot below represents the ages of 12 people.

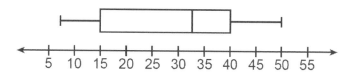

    What percentage of these people are age 15 or older?
        (1) 25   (3) 75
        (2) 35   (4) 85

9. The box-and-whisker plot below represents the results of tests scores in a math class.

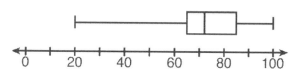

    What do the scores 65, 85, and 100 represent?
        (1) Q_1, median, Q_3 (3) median, Q_1, maximum
        (2) Q_1, Q_3, maximum   (4) minimum, median, maximum

10. The box-and-whisker plot below represents a set of grades in a college statistics class.

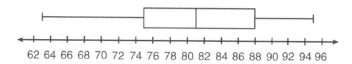

    Which interval contains exactly 50% of the grades?
        (1) 63-88   (3) 75-81
        (2) 63-95   (4) 75-88

207

11. The box-and-whisker plot shown below represents the number of magazine subscriptions sold by members of a club.

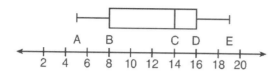

Which statistical measures do points B, D, and E represent, respectively?
   (1) minimum, median, maximum
   (2) first quartile, median, third quartile
   (3) first quartile, third quartile, maximum
   (4) median, third quartile, maximum

Constructed Response

12. The test scores from Mrs. Gray's math class are shown below.
    72, 73, 66, 71, 82, 85, 95, 85, 86, 89, 91, 92

   Construct a box-and-whisker plot to display these data.

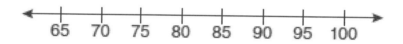

13. The number of songs fifteen students have on their MP3 players is:
    120, 124, 132, 145, 200, 255, 260, 292, 308, 314, 342, 407, 421, 435, 452

   State the values of the minimum, 1st quartile, median, 3rd quartile, and maximum. Using these values, construct a box-and-whisker plot using an appropriate scale on the line below.

14. Using the line provided, construct a box-and-whisker plot for the 12 scores below.
    26, 32, 19, 65, 57, 16, 28, 42, 40, 21, 38, 10

   Determine the number of scores that lie above the 75th percentile.

15. During the last 15 years of his baseball career, Andrew hit the following number of home runs each season.

35, 24, 32, 36, 40, 32, 40, 38, 36, 33, 11, 20, 19, 22, 8

State and label the values of the minimum, 1st quartile, median, 3rd quartile, and maximum.

Using the line below, construct a box-and-whisker plot for this set of data.

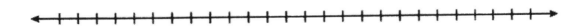

16. Robin collected data on the number of hours she watched television on Sunday through Thursday nights for a period of 3 weeks. The data are shown in the table below.

	Sun	Mon	Tues	Wed	Thurs
Week 1	4	3	3.5	2	2
Week 2	4.5	5	2.5	3	1.5
Week 3	4	3	1	1.5	2.5

Using an appropriate scale on the number line below, construct a box plot for the 15 values.

X.   BIVARIATE DATA

Two-Way Frequency Tables

Key Terms and Concepts

A **two-way table** (or *pivot table*) shows frequencies for bivariate data.

For example: A bakery holds a taste test in which participants select their favorite cup cake icing flavor. The two-way table below shows the results for 50 adults - 20 women and 30 men. In this survey, only 2 out of 20 women preferred vanilla, but 16 out of the 30 men chose vanilla.

	Vanilla	Chocolate	Strawberry	Total
Women	2	10	8	20
Men	16	6	8	30
Total	18	16	16	50

Entries in the body of the table are called **joint frequencies**. Entries in the "Total" row and "Total" column are called **marginal frequencies**. The marginal frequencies are the sums of the joint frequencies on that row or column of the table. The **grand total** in the lower right hand cell is the total number of data points. It should equal the sum of each set of marginal frequencies.

Sometimes we may prefer to show ratios (in *percent*, *decimal*, or *fraction* format). In this case, each entry, called a **relative frequency**, is a ratio of the frequency for that cell to the total number of data points. The total percentage written in the lower right hand cell is always 100%.

For example: We could have represented the above table using relative frequencies, below.

	Vanilla	Chocolate	Strawberry	Total
Women	4%	20%	16%	40%
Men	32%	12%	16%	60%
Total	36%	32%	32%	100%

Two-way tables help us to find conditional relative frequencies. A **conditional relative frequency** is one that is calculated given that a certain condition (row or column) is true.

For example: Using the relative frequency table above, we could say that among *women*, 10% prefer vanilla (4% divided by the 40% who are women). The given condition, "among women," restricts us to the top row. Or, we could say that among those who prefer *strawberry*, half are women (16% divided by 32%). In this case, the given condition restricts us to the column labelled "Strawberry."

Model Problem

A public opinion survey explored the relationship between age and support for increasing the minimum wage. The results are summarized in the two-way table below.

	For	Against	No opinion	Total
21 – 40	25	20	5	50
41 – 60	20	35	20	75
Over 60	55	15	5	75
Total	100	70	30	200

In the 21 to 40 age group, what percent supports increasing the minimum wage?

Solution: 50%

Explanation: A total of 50 people in the 21 to 40 age group were surveyed. Of those, 25 were "for" increasing the minimum wage. Therefore, 50% (25 ÷ 50) of the respondents in this age group supported the increase.

Practice Problems

1. In a survey of ninth and tenth grade students, participants were asked what grade they were in and whether they planned to watch the Super Bowl. Results are shown in the table below. What percent of the students are undecided? What percent of the ninth graders are watching? Round your answers to the *nearest tenth of a percent*.

	Watching	Not Watching	Undecided	Total
8th Grade	25	20	8	53
9th Grade	31	22	7	60
Total	56	42	15	113

2. The first table shows the number of books sold at a library sale. Complete this *joint frequency table* by writing the *marginal frequencies* in the blank cells. Then, using the same data, create an equivalent two-way *relative frequency table* in the second table below.

	Fiction	Nonfiction	Total
Hardcover	28	52	
Paperback	94	36	
Total			

	Fiction	Nonfiction	Total
Hardcover			
Paperback			
Total			

3. You go to a dance and help clean up afterwards. To help, you collect the soda cans, Coca-Cola and Sprite, and organize them. Some cans were on the table and some were in the garbage. 72 total cans were found. 42 total cans were found in the garbage and 50 total cans were Coca-Cola. 14 Sprite cans were found on the table. From the given information, complete the two-way joint frequency table below.

	Coca-Cola	Sprite	Total
Table			
Garbage			
Total			

Scatter Plots

Key Terms and Concepts

A **scatter plot** is a graph used to plot pairs of bivariate data values in a coordinate plane. They are often used for gathering experimental data to determine whether a correlation exists between the two variables. The **independent** variable is represented by **x-values** in a horizontal axis and the **dependent** variable is represented by **y-values** in a vertical axis. Only points are plotted; the points are not connected by lines.

The bivariate data can be written using a table. The *x*-values are always written first (the top row in a horizontal table or the left column in a vertical table), followed by the corresponding *y*-values. Each pair of values in the table can be plotted as a single point.

For example: Park administrators use a table to keep track of daily temperatures and the number of daily visitors to the beach. The points are then plotted on the scatter plot below, using the temperatures as *x*-values and visitors as *y*-values. The first data column is plotted as the point (84, 225) on the scatter plot.

Temp (x)	84	86	82	87	86	92	88	89	94	96	94
Visitors (y)	225	350	100	125	300	450	455	525	600	565	510

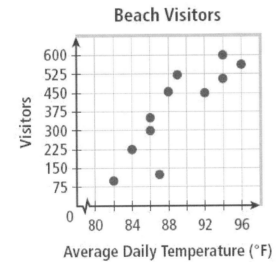

Each axis must include **labels** in equal intervals to cover the range of *x* or *y* values in the data, and should include an axis **title** describing what the axis labels represent.

For example: In the scatter plot above, the *x*-axis is labeled from 80 to 96 in intervals of 4, where each grid square is 2 units wide. To save room, you may omit labels between 0 and the first tick by using a ⌇ symbol, as shown between 0 and 80 on the *x*-axis. To cover values as high as 600 visitors, intervals of 75 visitors are used on the *y*-axis.

To enter bivariate data into the calculator:

1. Press [STAT] Edit... [ENTER].
2. If any values already appear in the L1 or L2 columns, select the column heading and press [CLEAR] [ENTER].
3. Enter the x values into the L1 column and the corresponding y values into the L2 column.

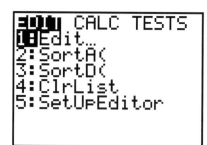

Notes: If any columns other than L1 or L2 appear in the table, you can delete them by selecting the column heading and pressing [DEL]. Also, if either the L1 or L2 column is missing, you can add it by selecting an empty column heading and then pressing [2nd] L1 [ENTER] or [2nd] L2 [ENTER].

To view a scatter plot on the graphing calculator:

1. Enter the x and y values into L1 and L2 as described above.
2. Press [2nd] [STAT PLOT] Plot1 [ENTER]. Select On [ENTER]. Be sure the Type is set at the first option ⸳⸱ , Xlist is L1 and Ylist is L2.
3. Press [ZOOM] ZoomStat [ENTER].

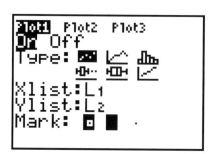

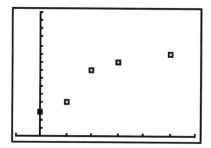

Note: When you are finished viewing the scatter plot, you may turn it off by entering [2nd] [STAT PLOT] Plot1 [ENTER] Off [ENTER].

Model Problem

A teacher records how many hours her students studied during the week leading up to their Regents exam and the scores they received. The data is shown in the following table. Create a scatter plot for the data.

Hours	3	5	2	6	7	1	2	7	1	7
Score	80	90	75	80	90	50	65	85	40	100

Solution:

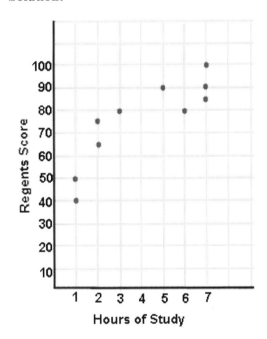

Explanation of Steps:

(A) Draw a grid. Label the x-axis using equal intervals covering the range of x values in the table *[1 through 7, with intervals of 1].*
Label the y-axis similarly to cover the range of y values *[10 through 100, with intervals of 10].*
Add appropriate axis titles.

(B) Plot each pair of data values as a point on the grid *[the first point in the table is (3, 80)].*

Enter the table into the calculator and create a scatter plot according to the instructions at the top of this page.

Practice Problems

1. The table shows the height (in inches) and the weight (in pounds) of five starters on a high school basketball team. Create the corresponding scatter plot.

Height (x)	67	72	77	74	69
Weight (y)	155	220	240	195	175

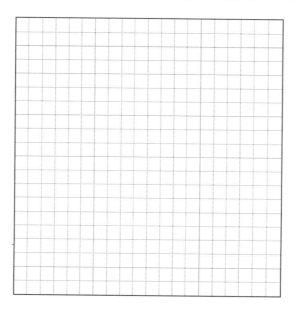

2. The following table lists weights (in hundreds of pounds) and highway fuel usage rates (in mpg) for a sample of domestic new cars. Create the corresponding scatter plot.

Weight (x)	29	35	28	44	25	34	30	33	28	24
Fuel (y)	31	27	29	25	31	29	28	28	28	33

REGENTS QUESTIONS

Multiple Choice

1. For 10 days, Romero kept a record of the number of hours he spent listening to music. The information is shown in the table below.

Day	1	2	3	4	5	6	7	8	9	10
Hours	9	3	2	6	8	6	10	4	5	2

Which scatter plot shows Romero's data graphically?

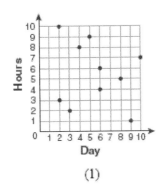

(1)

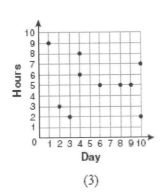

(3)

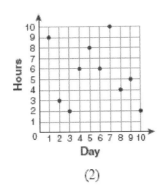

(2)

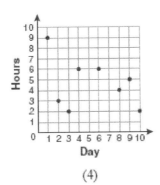

(4)

2. The school store did a study comparing the cost of a sweatshirt with the number of sweatshirts sold. The price was changed several times and the numbers of sweatshirts sold were recorded. The data are shown in the table below.

Cost of Sweatshirt	$10	$25	$15	$20	$5
Number Sold	9	6	15	11	14

Which scatter plot represents the data?

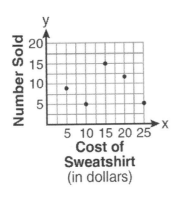

(1)

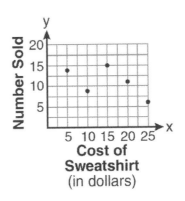

(3)

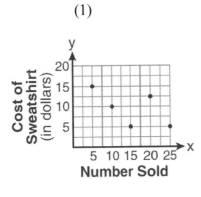

(2)

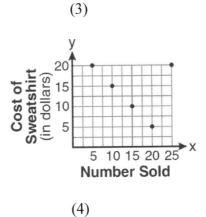

(4)

3. The maximum height and speed of various roller coasters in North America are shown in the table below.

Maximum Speed, in mph, (x)	45	50	54	60	65	70
Maximum Height, in feet, (y)	63	80	105	118	141	107

Which graph represents a correct scatter plot of the data?

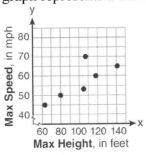

(1)

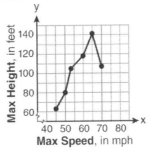

(3)

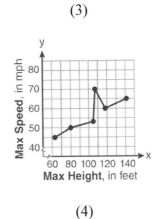

(2)

(4)

Correlation and Causality

Key Terms and Concepts

In a set of bivariate data, **correlation** (or *association*) is a statistical relationship between the two variables.

If the two variables are x and y, the data is said to have a **positive correlation** if, as x increases, y also increases, and as x decreases, y also decreases.

For example: As the temperature *increases*, *more* ice cream cones are sold. As the temperature *decreases*, *less* ice cream cones are sold. This shows a positive correlation.

There is a **negative correlation** when, as x increases, y decreases, and as x decreases, y increases.

For example: As the temperature *increases*, *less* cups of hot chocolate are sold. As the temperature *decreases*, *more* cups of hot chocolate are sold. This shows a negative correlation.

In some cases, a change in one variable is directly responsible for the change in the second variable. These are called **causal** relationships.

For example: When boiling a pot of water, the relationship between the time spent on the stove and the temperature of the water is causal.

However, the existence of a correlation does not necessarily mean that the relationship is causal. For example, a study of middle school students found a positive correlation between shoe sizes and reading comprehension scores. Clearly, a larger shoe size does not cause an increase in reading comprehension, or vice-versa, so this is *not a causal relationship*. In fact, a missing factor in this research is the age of the students: older children tend to have larger feet, and also tend to have higher reading scores.

Almost all real life relationships would have some hidden factors, but for our purposes we will define a "cause" as a primary factor responsible.

For example: In the above example about boiling water, a number of other factors could be partially responsible for the rise in water temperature. But certainly, placing a pot of water on a hot stove for a period of time will directly cause the temperature of the water to increase.

Model Problem

A number of children between the ages of 5 and 15 are measured for height. How would you describe the relationship between the ages and heights? Is there a correlation, and if so, is it positive or negative? Is it a causal relationship?

Solution:

Children grow as they get older, so there is a positive correlation and a causal relationship.

Explanation of Steps:

If an increase or decrease in the first variable *causes* a change in the second variable, there is a causal relationship. *[Yes, there are other hidden factors, such as nutrition, genetics, etc., but it safe to say that aging is a primary cause of growth within this age group.]* If an increase in one variable results in an increase in the other, it is a positive correlation, but if an increase in one variable results in a decrease in the other, it is a negative correlation.

220

Practice Problems

1. Identify the correlation you would expect to see (*positive*, *negative*, or *none*) between each pair of data sets. Explain.

    a) children's ages and their weights

    b) the volume of water poured into a container and the amount of empty space left in the container

    c) a person's shoe size and the length of the person's hair

    d) the outside temperature and the number of people at the beach

2. For each research finding below, (a) determine whether there is a positive, negative, or no correlation; (b) decide if there is a causal relationship; and (c) if a correlation is not causal, state what missing factors or variables, if any, may be the cause of the results.

    A. As the volume of air in a balloon increases, its diameter increases as well.

    B. As the number of workers increases, the number of days required to complete a job decreases.

    C. The more firefighters sent to a fire, the longer it takes to put out the fire.

    D. Over the past few centuries, the number of pirates worldwide has decreased while the level of CO_2 in the atmosphere has increased.

REGENTS QUESTIONS

Multiple Choice

1. Which situation describes a correlation that is *not* a causal relationship?
   (1) The rooster crows, and the sun rises.
   (2) The more miles driven, the more gasoline needed.
   (3) The more powerful the microwave, the faster the food cooks.
   (4) The faster the pace of a runner, the quicker the runner finishes.

2. Which relationship can best be described as causal?
   (1) height and intelligence
   (2) shoe size and running speed
   (3) number of correct answers on a test and test score
   (4) number of students in a class and number of students with brown hair

3. Which situation describes a correlation that is *not* a causal relationship?
   (1) the length of the edge of a cube and the volume of the cube
   (2) the distance traveled and the time spent driving
   (3) the age of a child and the number of siblings the child has
   (4) the number of classes taught in a school and the number of teachers employed

4. Which phrase best describes the relationship between the number of miles driven and the amount of gasoline used?
   (1) causal, but not correlated
   (2) correlated, but not causal
   (3) both correlated and causal
   (4) neither correlated nor causal

5. A study showed that a decrease in the cost of carrots led to an increase in the number of carrots sold. Which statement best describes this relationship?
   (1) positive correlation and a causal relationship
   (2) negative correlation and a causal relationship
   (3) positive correlation and not a causal relationship
   (4) negative correlation and not a causal relationship

6. Which situation does *not* describe a causal relationship?
   (1) The higher the volume on a radio, the louder the sound will be.
   (2) The faster a student types a research paper, the more pages the paper will have.
   (3) The shorter the distance driven, the less gasoline that will be used.
   (4) The slower the pace of a runner, the longer it will take the runner to finish the race.

7. Which situation describes a negative correlation?
   - (1) the amount of gas left in a car's tank and the amount of gas used from it
   - (2) the number of gallons of gas purchased and the amount paid for the gas
   - (3) the size of a car's gas tank and the number of gallons it holds
   - (4) the number of miles driven and the amount of gas used

8. Which situation describes a correlation that is *not* a causal relationship?
   - (1) the number of miles walked and the total Calories burned
   - (2) the population of a country and the census taken every ten years
   - (3) the number of hours a TV is on and the amount of electricity used
   - (4) the speed of a car and the number of hours it takes to travel a given distance

Identifying Correlation in Scatter Plots

Key Terms and Concepts

Although the points in a scatter plot may not be collinear (i.e., they may not lie in a straight line), one can visually determine if the **correlation** between the variables is positive or negative, or if no discernible correlation exists.

As with slope, the variables have a **positive correlation** if, as the *x*-values (*independent variable*) increase, so do the *y*-values (*dependent variable*) tend to increase. The variables have a **negative correlation** if the general tendency is that, as the *x*-values increase, the *y*-values decrease. It's also possible that the points show no such tendencies, in which case there is **no correlation**.

For example: The following diagrams show that (a) there is *no correlation* between a person's arm length and his or her results on an exam; (b) there is a *positive correlation* between the time a person spends revising the exam essay and the results on the exam; and (c) there is a *negative correlation* between the number of absences from school and the exam results.

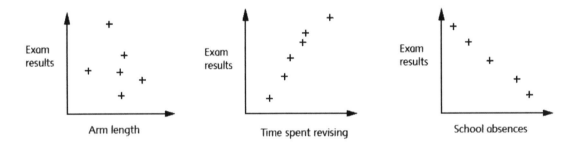

A correlation is considered **strong** if the points closely approximate a straight line (or curve).

For example: The first graph below shows a stronger positive correlation than the second.

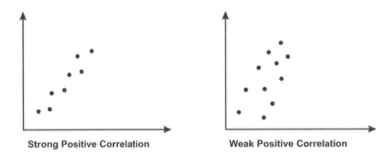

Model Problem

Which diagram shows a negative correlation?

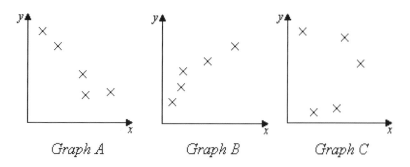

| Graph A | Graph B | Graph C |

Solution:

    Graph A

Explanation of Steps:

    A graph has a negative correlation if, as the x-values increase, the y-values tend to decrease. *[Graph B shows a positive correlation and Graph C shows no correlation.]*

Practice Problems

1. Which diagram shows the strongest positive correlation?

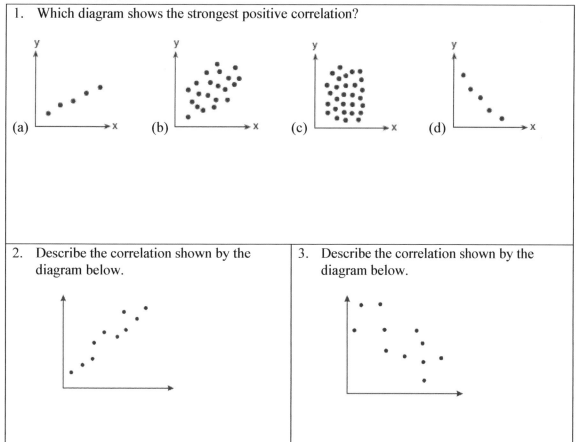

| 2. Describe the correlation shown by the diagram below. | 3. Describe the correlation shown by the diagram below. |

REGENTS QUESTIONS

Multiple Choice

1. There is a negative correlation between the number of hours a student watches television and his or her social studies test score. Which scatter plot below displays this correlation?

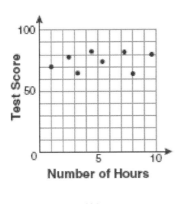

(1)

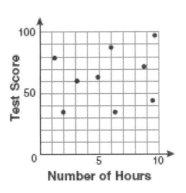

(3)

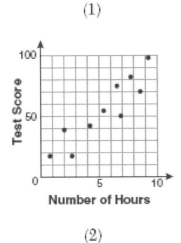

(2)

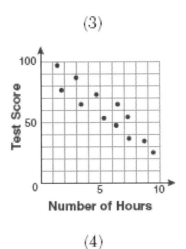

(4)

2. Which scatter plot shows the relationship between x and y if x represents a student score on a test and y represents the number of incorrect answers a student received on the same test?

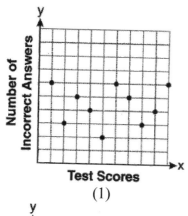

(1)

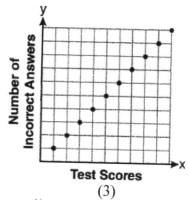

(3)

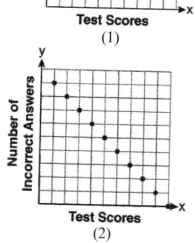

(2)

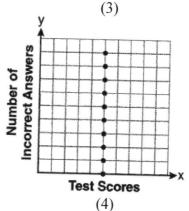

(4)

3. What is the relationship between the independent and dependent variables in the scatter plot shown below?

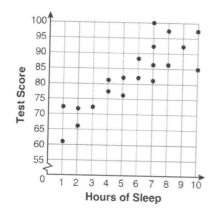

(1) undefined correlation

(2) negative correlation

(3) positive correlation

(4) no correlation

227

4. The scatter plot below represents the relationship between the number of peanuts a student eats and the student's bowling score.

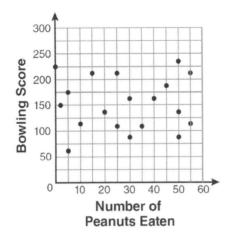

Which conclusion about the scatter plot is valid?
   (1) There is almost no relationship between eating peanuts and bowling score.
   (2) Students who eat more peanuts have higher bowling scores.
   (3) Students who eat more peanuts have lower bowling scores.
   (4) No bowlers eat peanuts.

5. A set of data is graphed on the scatter plot below.

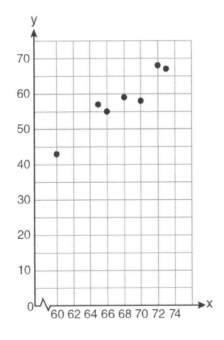

This scatter plot shows
   (1) no correlation   (3) negative correlation
   (2) positive correlation   (4) undefined correlation

6. The scatter plot shown below represents a relationship between x and y.

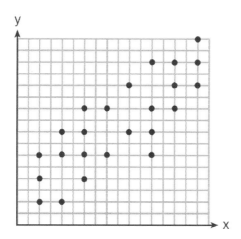

This type of relationship is
    (1) a positive correlation   (3) a zero correlation
    (2) a negative correlation   (4) not able to be determined

7. The number of hours spent on math homework during one week and the math exam grades for eleven students in Ms. Smith's algebra class are plotted below.

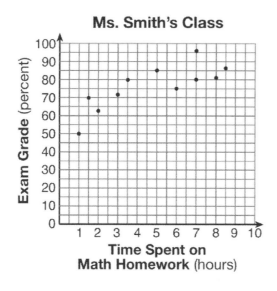

Based on the plotted data, what is the correlation between the time spent on homework and the exam grade?
    (1) positive   (3) no correlation
    (2) negative   (4) cannot be determined

Lines of Fit

Key Terms and Concepts

When a scatter plot shows a linear correlation, a line which approximates this relationship is called a **line of fit** (or *trend line*).

For example: For the following scatter plot, a line of fit may be drawn as shown. Because the graph shows a positive correlation, the line will have a positive slope.

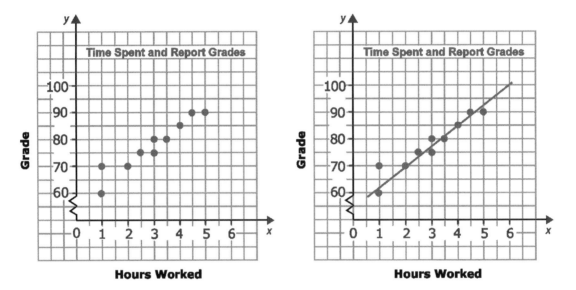

We can create a **line of fit** and find its equation by one of two methods: either by *drawing* an appropriate line by hand (using a straight edge) or by using the *calculator*. The calculator finds the **linear regression** (also known as the **least-squares line**), which passes through the mean point $(\bar{x}, \bar{y})$, where $\bar{x}$ is the mean of the x's and $\bar{y}$ is the mean of the y's. This line minimizes the combined distances of the points from the line, and so it is usually called a **line of best fit**. (More accurately, it minimizes the sum of the squared residuals; we will learn about residuals in the next section.)

Drawing a Line of Fit:
When drawing a straight line, we try to have the points lie as close to the line as possible, and preferably with as many points above the line as below it.
We can then determine its equation by finding two points which appear to lie directly on the line. Try not to pick points that are too close to each other. From these two points, we can determine a slope and y intercept.

For example: The line above appears to run through points (2, 70) and (4, 85).

Using these points, the slope $m = \dfrac{85 - 70}{4 - 2} = \dfrac{15}{2} = 7.5$.

Substituting point (2, 70) for x and y and the slope 7.5 for m in the general equation $y = mx + b$, we can find b:
$$70 = 7.5(2) + b \ \rightarrow \ 70 = 15 + b \ \rightarrow \ b = 55$$
Therefore, the equation for this line of fit is $y = 7.5x + 55$.

Using the Calculator to Find a Line of Best Fit:
1. Enter the *x* and *y* coordinates of the points as $\boxed{\text{L1}}$ and $\boxed{\text{L2}}$ (see page 214).
2. Press $\boxed{\text{STAT}}$ $\boxed{\text{CALC}}$ $\boxed{\text{LinReg(ax+b)}}$.
3. On the next screen prompt, "LinReg(ax+b)", press $\boxed{\text{VARS}}$ $\boxed{\text{Y-VARS}}$ $\boxed{\text{Function}}$ $\boxed{\text{Y1}}$ $\boxed{\text{ENTER}}$. The equation is stored in Y1.
4. The screen will show the equation y = ax + b along with the values of *a* (the slope) and *b* (the y-intercept).
5. To view the resulting line of best fit, press $\boxed{\text{ZOOM}}$ $\boxed{\text{ZoomStat}}$.
6. To see the equation of the line, press $\boxed{\text{Y=}}$.

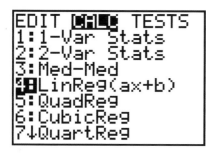

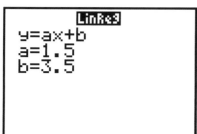
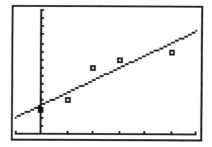

For example: If we enter the points as they appear in the graph on the previous page, the calculator will create the equation Y1=6.651718984X+57.877429. Rounding to the *nearest tenth*, the equation would be $y = 6.7x + 57.9$.

Note: Statisticians often use the form LinReg(**a+bx**), where *a* is the y-intercept and *b* is the slope, instead of LinReg(**ax+b**). They simply switch the terms. Since we only recently mastered the slope-intercept form as $y = mx + b$, I would recommend using LinReg(ax+b) for now.

The line of fit helps us to **predict** values not included in the original data. We can **extrapolate** about data that is outside (but near) the range of given *x*-values, or **interpolate** about data that is within the range of *x*-values but not already included in the data.

For example: Using the graph above, if a student submits a project on which he has worked 1.5 hours, we can predict his grade by looking at a line of fit or by using its equation. This would be *interpolation*, since 1.5 is within the range of *x* values. Either of the following predictions would be acceptable:
a) Substituting 1.5 for *x* in the equation of the drawn line, we get
$y = 7.5(1.5) + 55 = 66.25 \approx 66$.
b) Using the equation of the calculator's linear regression, we get
$y = 6.7(1.5) + 57.9 = 67.95 \approx 68$.

We could also have the calculator find this value by entering 1.5 $\boxed{\text{STO▶}}$ $\boxed{\text{ALPHA}}$ [X] $\boxed{\text{ENTER}}$ and then $\boxed{\text{VARS}}$ $\boxed{\text{Y-VARS}}$ $\boxed{\text{Function}}$ $\boxed{\text{Y1}}$ $\boxed{\text{ENTER}}$.

Model Problem

Given the data table below, create a scatter plot. Draw a reasonable line of fit and state its equation. Use the equation to extrapolate the next y value for an x value of 7.

x	0	1	2	3	4	5	6
y	2	4.5	9	11	13	18	19.5

Solution:

(A)

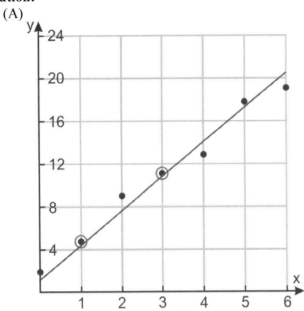

(B) From points (1, 4.5) and (3, 11), $m = \dfrac{11-4.5}{3-1} = \dfrac{6.5}{2} = 3.25$.

(C) Using (3, 11) and $m = 3.25$: $y = mx + b$ → $11 = 3.25(3) + b$ → $b = 1.25$

(D) Equation for the line of fit is $y = 3.25x + 1.25$

(E) For an x value of 7, $y = 3.25(7) + 1.25 = 24$

Explanation of Steps:

(A) Draw the scatter plot and a reasonable line of fit, where about as many points are above the line as below it. For our purposes, any reasonable line will be acceptable.

(B) Find two points that appear to lie on the line or are closest to it. Calculate the slope of the line between these two points using the slope formula.

(C) Use one of the points and the slope to substitute for x, y, and m in the general equation $y = mx + b$, and solve for b.

(D) Write the equation of the line.

(E) To extrapolate (or interpolate), use the new given value and substitute it into the equation to find the value of the other variable. *[Substitute 7 for x, and find y = 24.]*

Note: The equation of the calculator's regression line for this data would round to $y = 3x + 2$. It would start near the point (0,2) and have a slightly smaller slope than the drawn line. Using this line of best fit, we would extrapolate $y = 3(7) + 2 = 23$. Either answer is acceptable.

Practice Problems

1. The following chart shows students' typing speeds in words per minute (wpm) after a certain number of weeks of practice.

    (a) Based on the line of fit shown, approximately how fast would you expect a student to type after 8 weeks of practice?

    (b) By how many words per minute, approximately, can an employee expect to increase her or his speed for each additional week of practice?

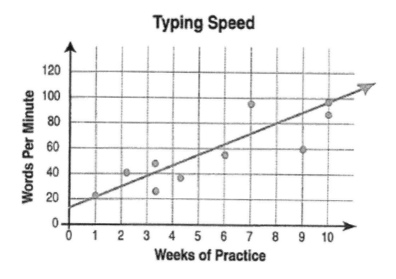

2. The scatter plots below display the same data about the ages of eight health club members and their heart rates during exercises. Which line is a better fit for the data? Explain your reasoning.

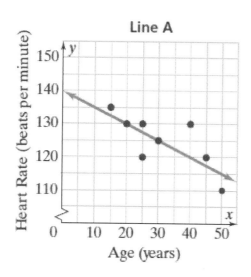
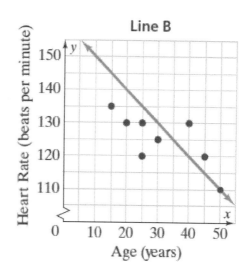

3. The local ice cream shop keeps track of daily sales (in dollars) and the temperature (in Celsius) on that day. On the scatter plot below, draw a line of fit.

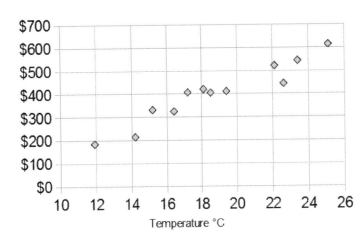

Temperature °C

4. Write an equation for the line of fit shown in the scatter plot below.

Months Employed

5. The data below shows hours spent researching the stock market per week and the percent gain for an investor.

Hours	6	8	10	12	14	16	18
% Gain	17	20.5	26.5	29	32.5	37.5	41

Find an equation of the line of best fit for gain with respect to hours of study.

6. A random sample of graduates from a particular college program reported their ages and incomes in response to a survey. Each point on the scatter plot below represents the age and income of a different graduate. Of the following equations, which best fits the data?

   (a) $y = -1,000x + 15,000$ (c) $y = 1,000x + 15,000$

   (b) $y = 1,000x$ (d) $y = 10,000x + 15,000$

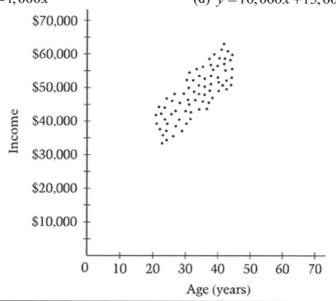

7. Based on the data in the scatter plot of the previous question, predictions can be made about the income of a 35 year old and the income of a 55 year old. For which age is the prediction more likely to be accurate? Justify your answer.

8. The table shows the temperature, t degrees Fahrenheit, displayed on an oven while it was heating as a function of the amount of time, s seconds, since it was turned on. Create a scatter plot for this data. Find the equation for its line of best fit.

s	t
31	175
61	200
104	225
158	250
202	275
250	300
285	325
327	350
380	375
428	400

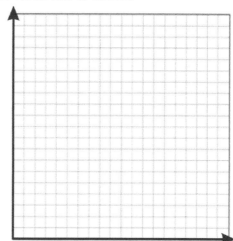

235

REGENTS QUESTIONS

Multiple Choice

1. Which equation most closely represents the line of best fit for the scatter plot below?

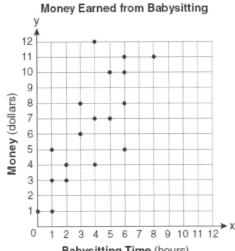

Money Earned from Babysitting

(1) $y = x$

(2) $y = \dfrac{2}{3}x + 1$

(3) $y = \dfrac{3}{2}x + 4$

(4) $y = \dfrac{3}{2}x + 1$

2. The number of hours spent on math homework each week and the final exam grades for twelve students in Mr. Dylan's algebra class are plotted below.

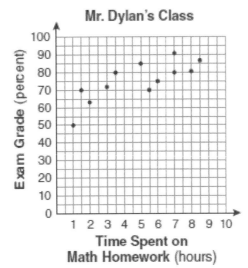

Mr. Dylan's Class

Based on a line of best fit, which exam grade is the best prediction for a student who spends about 4 hours on math homework each week?

(1) 62   (3) 82

(2) 72   (4) 92

3. A scatter plot was constructed on the graph below and a line of best fit was drawn.

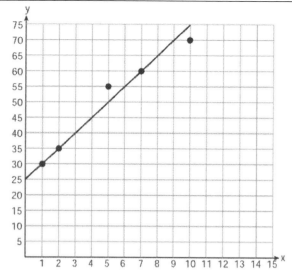

What is the equation of this line of best fit?

(1) $y = x + 5$

(2) $y = x + 25$

(3) $y = 5x + 5$

(4) $y = 5x + 25$

4. The scatter plot below shows the profit, by month, for a new company for the first year of operation. Kate drew a line of best fit, as shown in the diagram.

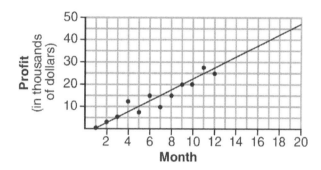

Using this line, what is the best estimate for profit in the 18th month?

(1) $35,000

(2) $37,750

(3) $42,500

(4) $45,000

5. Based on the line of best fit drawn below, which value could be expected for the data in June 2015?

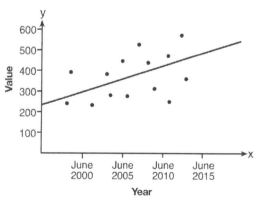

(1) 230

(2) 310

(3) 480

(4) 540

6. The graph below illustrates the number of acres used for farming in Smalltown, New York, over several years.

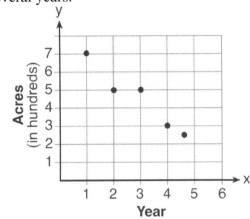

Using a line of best fit, approximately how many acres will be used for farming in the 5th year?

(1) 0   (3) 300
(2) 200   (4) 400

Constructed Response

7. The accompanying table shows the enrollment of a preschool from 1980 through 2000. Write a linear regression equation to model the data in the table.

Year (x)	Enrollment (y)
1980	14
1985	20
1990	22
1995	28
2000	37

8. The data table below shows water temperatures at various depths in an ocean.

Water Depth (x) (meters)	Temperature (y) (°C)
50	18
75	15
100	12
150	7
200	1

Write the linear regression equation for this set of data, rounding all values to the *nearest thousandth*. Using this equation, predict the temperature (°C), to the *nearest integer*, at a water depth of 255 meters.

9. In a mathematics class of ten students, the teacher wanted to determine how a homework grade influenced a student's performance on the subsequent test. The homework grade and subsequent test grade for each student are given in the accompanying table.

Homework Grade (x)	Test Grade (y)
94	98
95	94
92	95
87	89
82	85
80	78
75	73
65	67
50	45
20	40

a) Give the equation of the linear regression line for this set of data.
b) A new student comes to the class and earns a homework grade of 78. Based on the equation in part *a*, what grade would the teacher predict the student would receive on the subsequent test, to the *nearest integer*?

10. The table below shows the number of prom tickets sold over a ten-day period.

Prom Ticket Sales

Day (x)	1	2	5	7	10
Number of Prom Tickets Sold (y)	30	35	55	60	70

Plot these data points on the coordinate grid below. Use a consistent and appropriate scale. Draw a reasonable line of best fit and write its equation.

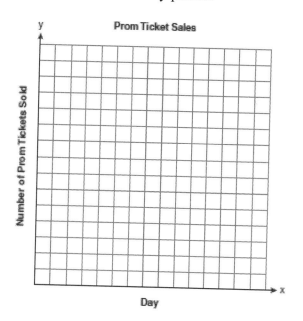

Prom Ticket Sales

11. Megan and Bryce opened a new store called the Donut Pit. Their goal is to reach a profit of $20,000 in their 18th month of business. The table and scatter plot below represent the profit, *P*, in thousands of dollars, that they made during the first 12 months.

t (months)	P (profit, in thousands of dollars)
1	3.0
2	2.5
3	4.0
4	5.0
5	6.5
6	5.5
7	7.0
8	6.0
9	7.5
10	7.0
11	9.0
12	9.5

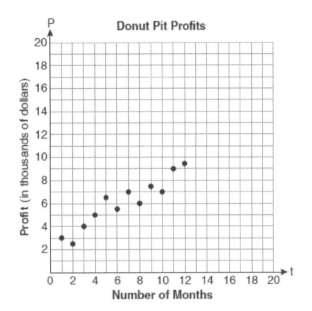

Draw a reasonable line of best fit. Using the line of best fit, predict whether Megan and Bryce will reach their goal in the 18th month of their business. Justify your answer.

240

Residuals and Correlation Coefficients

Key Terms and Concepts

The **residual** of a data point is a measure of how far the *y*-value of the point is from the line of best fit (that is, the predicted value). It is calculated as **actual y-value − predicted y-value**. Points *above* the line will have *positive* residuals; points *below* will have *negative* residuals.

For example: The point with the largest residual on the graph below, meaning the farthest point from the line of best fit, is (9,60).

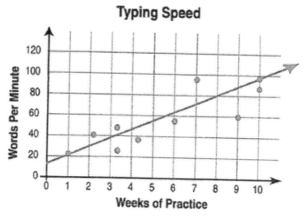

Suppose the equation for the line of best fit is $y = 8.7x + 12.3$.

To find the residual for point (9,60), substitute 9 for *x* in the equation, giving us $y = 8.7(9) + 12.3 = 90.6$, so the residual of (9,60) is $60 - 90.6 = -30.6$.

The **correlation coefficient** tells us the degree of correlation; in other words, how close the entire set of points is from the line. It is a value between −1 and 1, with *negative* values used for best-fit lines with *negative* slopes and *positive* values used for lines with *positive* slopes. A correlation coefficient of 0 means no correlation; a value close to 0 ($|r| < 0.3$) represents a weak correlation.

A value of −1 or 1 would represent data points that are collinear (all points are on the regression line); a value close to −1 or 1 ($|r| \geq 0.7$) represents a strong correlation.

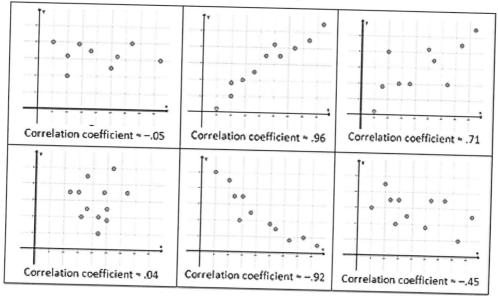

We can find the correlation coefficient on the calculator:

1. First, store the equation of the linear regression in Y1 (see page 231).
2. Press $\boxed{\text{STAT}}$ $\boxed{\text{TESTS}}$, arrow down to select $\boxed{\text{LinRegTTest}}$ and press $\boxed{\text{ENTER}}$.
3. Place the cursor next to $\boxed{\text{RegEQ}}$ and press $\boxed{\text{VARS}}$ $\boxed{\text{Y-VARS}}$ $\boxed{\text{Function}}$ $\boxed{\text{Y1}}$ $\boxed{\text{ENTER}}$.
4. Select $\boxed{\text{Calculate}}$ and press $\boxed{\text{ENTER}}$.
5. Arrow down to the end, where *r* gives the correlation coefficient.

```
EDIT CALC TESTS
0↑2-SampTInt…
A:1-PropZInt…
B:2-PropZInt…
C:X²-Test…
D:2-SampFTest…
E◼LinRegTTest…
F:ANOVA(
```

```
LinRegTTest
Xlist:L₁
Ylist:L₂
Freq:1
β & ρ:≠0 <0 >0
RegEQ:Y₁▨
Calculate
```

```
LinRegTTest
 y=a+bx
 β>0 and ρ>0
↑b=11.27586207
 s=20.81855108
 r²=.3471357865
 r=.5891823032
◼
```

Note: There is another way to see correlation coefficients on the calculator. If you turn Diagnostics on, then any time you calculate a linear regression, you will be told the correlation coefficient *r* along with the values of *a* and *b*. You can turn Diagnostics on by pressing $\boxed{\text{2nd}}$ [CATALOG], scrolling down to $\boxed{\text{DiagnosticOn}}$ and then pressing $\boxed{\text{ENTER}}$ twice.

```
CATALOG        ▣
 Degree
 DelVar
 DependAsk
 DependAuto
 det(
 DiagnosticOff
▶DiagnosticOn
```

Now, when you find a linear regression using LinReg, instead of showing the screen to the left below, the calculator will display a screen like the one to the right, including *r*.

```
LinReg
 y=ax+b
 a=10.5
 b=.1
```

```
LinReg
 y=ax+b
 a=10.5
 b=.1
 r²=.9983700081
 r=.9991846717
◼
```

Practice Problems

1. Based on a regression line, it is predicted that a 10 year old Toyota Corolla should cost $14,050. Its actual cost is $12,550. What is the residual?

2. Which of the following is the value of the correlation coefficient for the data set shown in the scatter plot below?
   (a) –0.95   (b) –0.24   (c) 0.83   (d) 1.00

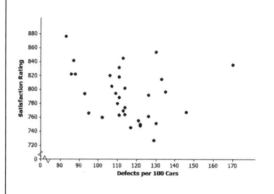

3. The correlation coefficients for the six scatter plots shown below are
   -0.85, -0.40, 0, 0.50, 0.90, 0.99
   Match each scatter plot with the correct correlation coefficient.

a.

b.

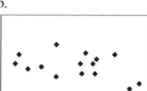

c.

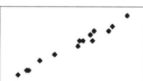

d.

e.

f.

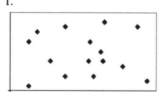

4. Use the calculator to find the correlation coefficient for the data below.

Exit #	2	39	48	57	67	75	91	110
Toll	1.50	2.50	3.00	3.50	4.25	4.50	5.50	6.50

5. The table below shows study time, in hours, and test scores based on a linear model for which $y = 8.8x + 58.4$ is the equation of a line of fit. Complete the table to show the predicted test score and residual for each point.

Study Time in Hours (x)	Test Score (y)	Predicted Test Score	Residual
0.5	63		
1	67		
1.5	72		
2	76		
2.5	80		
3	85		
3.5	89		

6. A linear regression equation to predict the number of points scored by an NFL team in one game based on the team's time of possession (in minutes) in that game is $y = 0.75x - 0.25$.

a) How many points would you predict a team to score if their time of possession was 22 minutes?

b) Say a team had 34 minutes of possession and scored 32 points. What is their residual?

c) A team had 28 minutes of possession and a residual of –0.75. How many points did they score?

REGENTS QUESTIONS

Multiple Choice

1. Which value of *r* represents data with a strong positive linear correlation between two variables?

   (1) 0.89   (3) 1.04
   (2) 0.34   (4) 0.01

2. Which value of *r* represents data with a strong negative linear correlation between two variables?

   (1) –1.07   (3) –0.14
   (2) –0.89   (4) 0.92

3. What could be the approximate value of the correlation coefficient for the accompanying scatter plot?

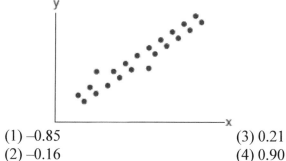

   (1) –0.85   (3) 0.21
   (2) –0.16   (4) 0.90

4. Which graph represents data used in a linear regression that produces a correlation coefficient closest to –1?

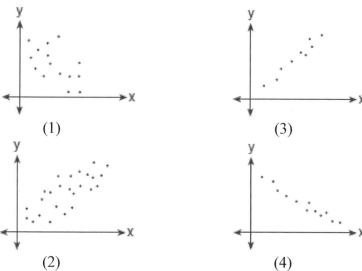

5. The relationship between t, a student's test scores, and d, the student's success in college, is modeled by the equation $d = 0.48t + 75.2$. Based on this linear regression model, the correlation coefficient could be

    (1) between –1 and 0   (3) equal to –1
    (2) between 0 and 1   (4) equal to 0

6. The relationship of a woman's shoe size and length of a woman's foot, in inches, is given in the accompanying table.

Woman's Shoe Size	5	6	7	8
Foot Length (in)	9.00	9.25	9.50	9.75

    The linear correlation coefficient for this relationship is

    (1) 1   (3) 0.5
    (2) –1   (4) 0

7. As shown in the table below, a person's target heart rate during exercise changes as the person gets older.

Age (years)	Target Heart Rate (beats per minute)
20	135
25	132
30	129
35	125
40	122
45	119
50	115

    Which value represents the linear correlation coefficient, rounded to the *nearest thousandth*, between a person's age, in years, and that person's target heart rate, in beats per minute?

    (1) –0.999   (3) 0.998
    (2) –0.664   (4) 1.503

8. What is the correlation coefficient of the linear fit of the data shown below, to the *nearest hundredth?*

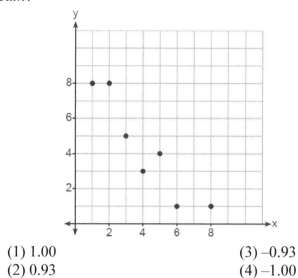

    (1) 1.00   (3) –0.93
    (2) 0.93   (4) –1.00

Constructed Response

9. Two different tests were designed to measure understanding of a topic. The two tests were given to ten students with the following results:

Test x	75	78	88	92	95	67	58	72	74	81
Test y	81	73	85	88	89	73	66	75	70	78

Construct a scatter plot for these scores, and then write an equation for the line of best fit (round slope and intercept to the *nearest hundredth*). Find the correlation coefficient. Predict the score, to the *nearest integer*, on test *y* for a student who scored 87 on test *x*.

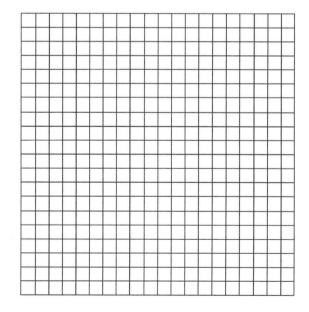

Residual Plots

Key Terms and Concepts

A **residual plot** is made by plotting the *x-values* on the *horizontal axis* and the corresponding *residuals* on the *vertical axis*.

A scatter plot with a line of best fit is shown below. Line segments are added to show the residual of each point. The length of each segment represents the point's residual. When plotted on a residual plot, the line of best fit is drawn as a horizontal line at 0, and the points are plotted using the same *x*-values along the horizontal axis but with the *labels on the vertical axis representing their residuals*.

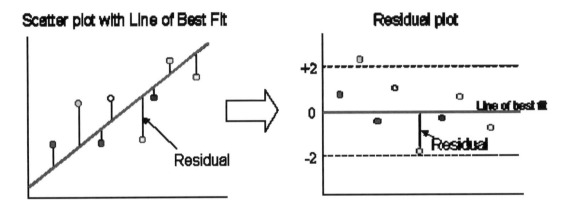

When creating a residual plot from a scatter plot, we can imagine residual line segments as strings tying the points to the regression line, and visualize rotating the regression line so that it becomes horizontal, while keeping the residual line segments (strings) vertical. For example:

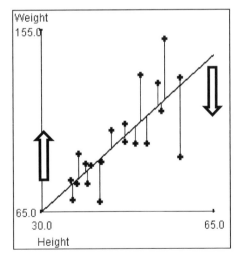
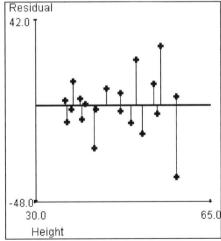

To create a residual plot on the calculator:
1. First, create a scatter plot for the data (see page 214) and store the equation of the linear regression in Y1 (see page 231).
2. Turn off Plot1 by pressing 2nd [STAT PLOT] Plot1 ENTER Off ENTER.
3. Select Plot2 ENTER On ENTER. Be sure the Type is set at the first option ⊡ and that Xlist is set to L1. For Ylist, press 2nd [LIST] NAMES RESID ENTER.
4. Press ZOOM ZoomStat ENTER.

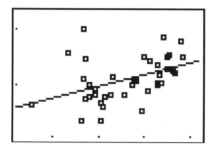 

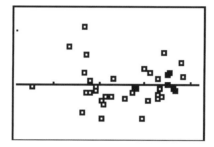

In a residual plot, the points should not show any pattern or curve. If they do, then the *linear* regression may not have been appropriate for the data. (In a future course, you may learn how to transform the regression equation into a *non-linear* model. For examples, you may use quadratic or exponential regression equations.)

For example: The first residual plot is shown on data for which a linear regression is appropriate. The residuals are more or less evenly distributed above and below the axis and show no particular pattern. The second residual plot does show a pattern, so a linear regression should not have been used. In other words, a *line* was not a best fit for this set of data.

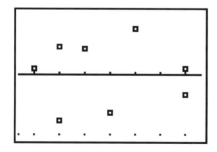 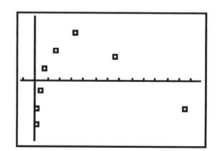

Practice Problems

1. Complete the following table using the equation $y = 0.5x$ as the line of fit to determine the predicted values. Round to the *nearest tenths*.

x	y	Predicted Value	Residual
5	3		
10	4		
15	9		
20	7		
25	13		
30	15		

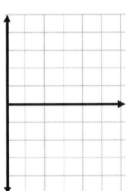

Then plot the residuals on the grid to the right.

Does the residual plot suggest a linear relationship? Explain.

2. Complete the following table using the equation $y = -0.4x + 16.3$ as the line of fit to determine the predicted values. Round to the *nearest tenths*.

x	y	Predicted Value	Residual
2	5		
4	15		
6	26		
8	23		
10	11		
12	3		

Then plot the residuals on the grid to the right.

Does the residual plot suggest a linear relationship? Explain.

3. Airlines charge different prices based on the distance of the flight being purchased. The table below shows the distances and prices of different flights for a given airline at an airport.

Destination	Distance (miles)	Airfare ($)	Predicted Price ($)	Residual
Atlanta	576	178		
Boston	370	138		
Chicago	612	94		
Dallas/Fort Worth	1,216	278		
Detroit	409	158		
Denver	1,502	258		
Miami	946	198		
New Orleans	998	188		
New York	189	98		
Orlando	787	179		
Pittsburgh	210	138		
St. Louis	737	98		

a) Find the equation for the line of best fit using the calculator.

b) Using this equation, calculate the predicted price for each flight and enter the predicted prices in the table above.
c) Calculate the residuals and enter them into the table above.
d) Using the grid below, graph the residual plot. Use your calculator to verify that your residual plot is correct.

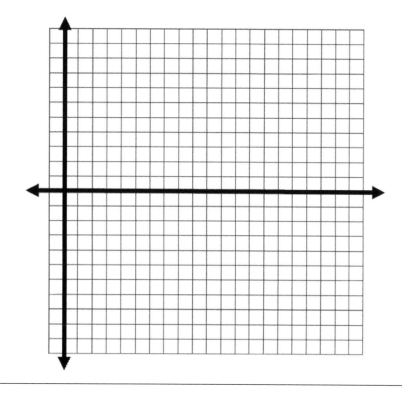

XI. INTRODUCTION TO FUNCTIONS

Determining if Relations are Functions

Key Terms and Concepts

An **ordered pair** is a pair of two values (*entries*) written in a certain order. The coordinates of a point on a graph are usually represented by an ordered pair in parentheses, with the *x*-coordinate (*abscissa*) written first and the *y*-coordinate (*ordinate*) written second, as in (3,–2).

A **relation** is a set of ordered pairs. It may be graphed as a set of points.

A **function** is a type of relation in which every first entry is mapped to exactly one second entry. In a function, no two ordered pairs can have the same first entries but different second entries. Represented as a graph, no two points may have the same *x*-value but different *y*-values.

Model Problem
Determine whether the relation, {(1,–1), (0,0), (1,1), (4,2)}, is a function or is *not* a function.

Solution:
It is *not* a function.

Explanation of Steps:
Look for any two ordered pairs that have the same *x*-coordinate but different *y*-coordinates *[(1,–1) and (1,1)]*. A relation is a function only if no such cases can be found.

Practice Problems

1. Which set of ordered pairs is *not* a function? (a) {(1,2), (3,4), (4,5), (5,6)} (b) {(3,1), (2,1), (1,2), (3,2)} (c) {(4,1), (5,1), (6,1), (7,1)} (d) {(0,0), (1,1), (2,2), (3,3)}	2. Which set of ordered pairs does *not* represent a function? (a) {(3,-2), (4,-3), (5,-4), (6,-5)} (b) {(3,-2), (3,-4), (4,-1), (4,-3)} (c) {(3,-2), (5,-2), (4,-2), (-1,-2)} (d) {(3,-2), (-2,3), (4,-1), (-1,4)}

3. The table to the right shows all the ordered pairs (*x, y*) that define a relation between the variables *x* and *y*. Is *y* a function of *x*? Justify your answer.

x	y
–2	3
–1	0
0	–1
1	0
2	3
3	8

REGENTS QUESTIONS

Multiple Choice

1. Which set of ordered pairs is *not* a function?
   - (1) {(3,1), (2,1), (1,2), (3,2)}
   - (2) {(4,1), (5,1), (6,1), (7,1)}
   - (3) {(1,2), (3,4), (4,5), (5,6)}
   - (4) {(0,0), (1,1), (2,2), (3,3)}

2. Which set of ordered pairs does *not* represent a function?
   - (1) {(3,–2), (–2,3), (4,–1), (–1,4)}
   - (2) {(3,–2), (3,–4), (4,–1), (4,–3)}
   - (3) {(3,–2), (4,–3), (5,–4), (6,–5)}
   - (4) {(3,–2), (5,–2), (4,–2), (–1,–2)}

3. Which relation is *not* a function?
   - (1) {(1,5), (2,6), (3,6), (4,7)}
   - (2) {(4,7), (2,1), (-3,6), (3,4)}
   - (3) {(-1,6), (1,3), (2,5), (1,7)}
   - (4) {(-1,2), (0,5), (5,0), (2-1)}

4. Which relation represents a function?
   - (1) {(0,3), (2,4), (0,6)}
   - (2) {(–7,5), (–7,1), (–10,3), (–4,3)}
   - (3) {(2,0), (6,2), (6,–2)}
   - (4) {(–6,5), (–3,2), (1,2), (6,5)}

5. Which relation is a function?
   - (1) $\left\{ \left(\frac{3}{4},0 \right),(0,1),\left(\frac{3}{4},2 \right) \right\}$
   - (2) $\left\{ (-2,2),\left(-\frac{1}{2},1 \right),(-2,4) \right\}$
   - (3) {(–1,4),(0,5),(0,4)}
   - (4) {(2,1),(4,3),(6,5)}

6. Which set of ordered pairs represents a function?
   - (1) {(0,4),(2,4),(2,5)}
   - (2) {(6,0),(5,0),(4,0)}
   - (3) {(4,1),(6,2),(6,3),(5,0)}
   - (4) {(0,4),(1,4),(0,5),(1,5)}

7. Which relation is *not* a function?
   - (1) {(2,4),(1,2),(0,0),(−1,2),(−2,4)}
   - (2) {(2,4),(1,1),(0,0),(−1,1),(−2,4)}
   - (3) {(2,2),(1,1),(0,0),(−1,1),(−2,2)}
   - (4) {(2,2),(1,1),(0,0),(1,−1),(2,−2)}

8. Which relation is a function?
   - (1) {(2,1), (3,1), (4,1), (5,1)}
   - (2) {(1,2), (1,3), (1,4), (1,5)}
   - (3) {(2,3), (3,2), (4,2), (2,4)}
   - (4) {(1,6), (2,8), (3,9), (3,12)}

Determining if Graphs Represent Functions

Key Terms and Concepts

Represented as a graph, a relation is a function only if no points have the same x-coordinates but different y-coordinates.

To test whether the graph of a relation is a function, we can use the **vertical line test**. If we can draw a vertical line that intersects the graph at two or more points, then these points have the same x-values but different y-values, and therefore the relation is not a function. It is a function only if it is *impossible* to draw a vertical line that would intersect the graph at multiple points.

Model Problem

Determine whether the graph below represents a function.

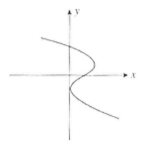

Solution:
The graph is *not* a function.

Explanation of Steps:
A graph is not a function if we are able to draw any vertical line that intersects the graph at two or more points. *[We can draw the vertical line below, for just one example. Since it crosses the graph in three points, this cannot be a function.]*

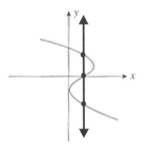

Practice Problems

1. Which of the following graphs is a function?

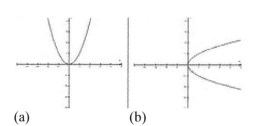

(a)   (b)

2. Which of the following graphs is a function?

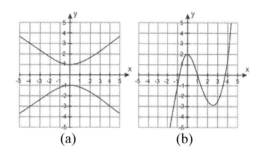

(a)   (b)

3. Which graph is *not* a function?

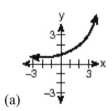   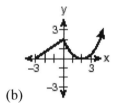   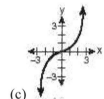   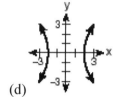

(a)   (b)   (c)   (d)

4. Which graph does *not* represent a function?

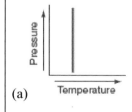   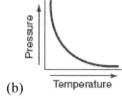   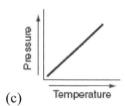   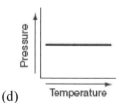

(a)   (b)   (c)   (d)

REGENTS QUESTIONS

Multiple Choice

1. Which graph represents a function?

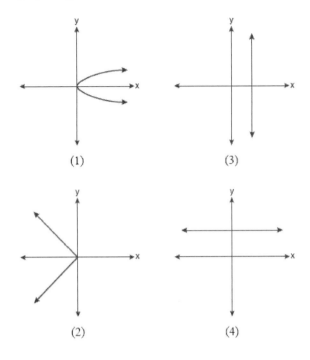

2. Which graph represents a function?

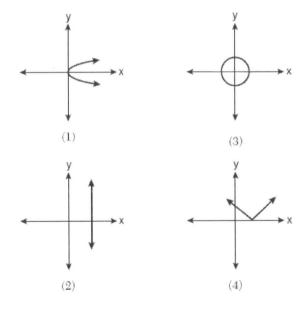

3. Which statement is true about the relation shown on the graph below?

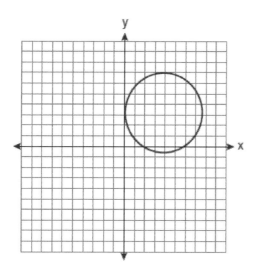

(1) It is a function because there exists one *x*-coordinate for each *y*-coordinate.
(2) It is a function because there exists one *y*-coordinate for each *x*-coordinate.
(3) It is *not* a function because there are multiple *y*-values for a given *x*-value.
(4) It is *not* a function because there are multiple *x*-values for a given *y*-value.

4. Which graph represents a function?

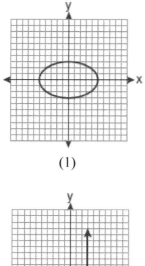

(1)

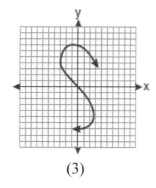

(3)

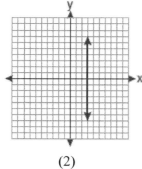

(2)

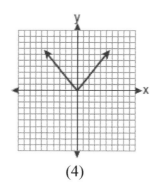

(4)

5. Which graph does *not* represent a function?

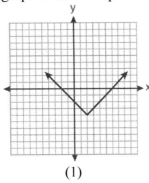

(1)

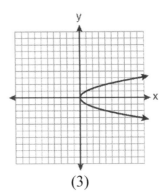

(3)

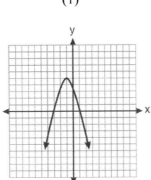

(2)

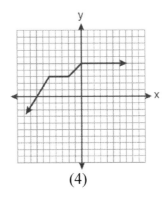

(4)

6. Which graph represents a function?

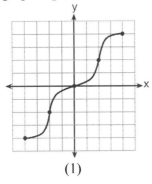

(1)

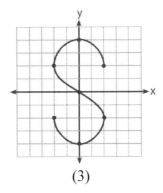

(3)

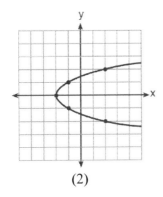

(2)

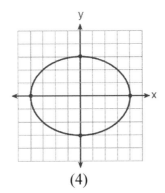

(4)

7. Which graph represents a function?

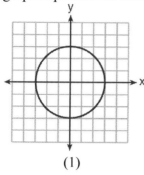

(1)

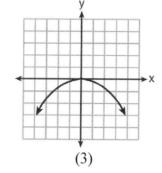

(3)

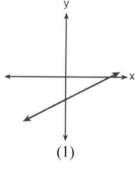

(2)

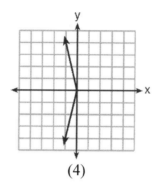

(4)

8. Which graph does *not* represent the graph of a function?

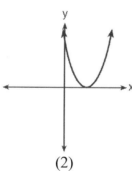

(1)

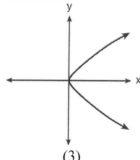

(3)

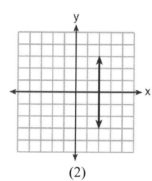

(2)

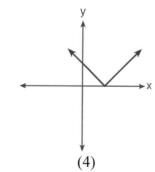

(4)

Function Notation, Domain and Range

Key Terms and Concepts

As we have seen, a function may be represented as a set of ordered pairs (x, y) such that each x is paired with a unique y. Often, the relationship between each x and its corresponding y can be represented by an algebraic expression.

For example: Considering the function represented by the expression $3x+5$, for each x-value, the corresponding y-value is 5 more than 3 times the x-value. Note that for each x-value, there is only one possible y-value.

A function is named using a letter (often f) followed by the expression's independent variable (often x) in parentheses, as in $f(x)$. This is read as "f of x". We could **define** the function by writing that it is equal to an expression in terms of the independent variable.

For examples: $f(x) = 3x+5$ $A(n) = n^2 - 2n + 1$

The set of possible x-values is called the **domain**, and the set of y-values that are produced by the function is called the **range**.

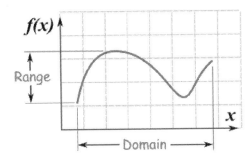

The function is like a machine in that for any given **input** value from the *domain*, the function produces a unique **output** value in the *range*.

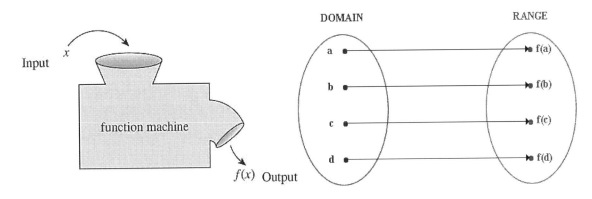

Restrictions on the domain:
If the domain is not specified, it is assumed to be the set of real numbers. However, the domain may be restricted to only certain intervals of the real numbers such as $x \geq 0$ or $-2 \leq x \leq 2$, or to *discrete* sets such as the set of integers, or even to *finite* sets such as $\{-2, -1, 0, 1, 2\}$. Or the domain may simply exclude certain values of x; for example, the domain for the function $f(x) = \frac{1}{x}$ for $x \neq 0$ excludes 0 from the domain because $\frac{1}{0}$ is undefined.

Restrictions on the domain may derive from the situation that the function models.

For examples: (a) If x represents the length of a side of a triangle, then $x > 0$ would be an appropriate domain for $f(x)$.

(b) If x represents a number of people, then *the set of whole numbers* would be an appropriate domain for $f(x)$.

(c) If x is the result of rolling a six-sided die, then the domain $\{1, 2, 3, 4, 5, 6\}$ would be appropriate for $f(x)$.

A **discrete domain** is a set of input values that consist of only certain numbers in an interval, whereas a **continuous domain** consists of all numbers in an interval.

For example: A domain of counting numbers from 1 to 5 would be discrete. A domain of real numbers from 1 to 5 would be continuous.

Practice Problems

1. State the range of the following function. $\{ (-1,2), (2, 51), (1, 3), (8, 22), (9, 51) \}$	2. What is an appropriate domain of the function $f(x) = \dfrac{1}{x}$?
3. Given $f(x) = x^2$ for all real numbers, x. (a) What is the range of this function? (b) Suppose we restrict the domain of this function to $-3 \le x \le 3$. How does this affect the range?	4. Suppose $g(n) = n + 1$ for the domain of whole numbers, n. Describe the range of this function.

5. Suppose n represents the number of multiple-choice questions answered correctly on a 20-question test. The function $f(n)$ represents the points earned on the test, where each question is worth 5 points with no partial credit.
   (a) Define the function $f(n)$.
   (b) What is an appropriate domain?
   (c) What is the range?

REGENTS QUESTIONS

Multiple Choice

1. Officials in a town use a function, C, to analyze traffic patterns. C(n) represents the rate of traffic through an intersection where n is the number of observed vehicles in a specified time interval. What would be the most appropriate domain for the function?

    (1) $\{...-2,-1,0,1,2,3,...\}$ (3) $\{0,\frac{1}{2},1,1\frac{1}{2},2,2\frac{1}{2}\}$

    (2) $\{-2,-1,0,1,2,3\}$ (4) $\{0,1,2,3,...\}$

2. If $f(x)=\frac{1}{3}x+9$, which statement is always true?

    (1) $f(x)<0$ (3) If $x<0$, then $f(x)<0$

    (2) $f(x)>0$ (4) If $x>0$, then $f(x)>0$

Constructed Response

3. The function *f* has a domain of $\{1, 3, 5, 7\}$ and a range of $\{2, 4, 6\}$.
   Could *f* be represented by $\{(1,2), (3,4), (5,6), (7,2)\}$? Justify your answer.

Function Graphs

Key Terms and Concepts

Function graphs: When a function $f(x)$ is graphed on a coordinate graph, the y-values in the range are the values produced by the function. For this reason, the y-axis is often labelled $f(x)$. For example: $f(x) = 3x + 5$ is graphed as the line $y = 3x + 5$. Therefore, $f(x) = y$.

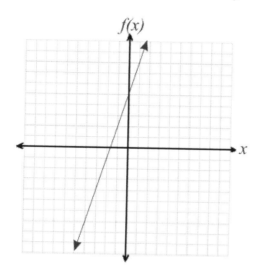

Linear Function
$y = mx + b$
$f(x) = mx + b$

A function's value, $f(x)$, for a given value of x can be determined by finding the y-value of a point (x,y) for the given x on a graph of the function.

For example: For the function $f(x) = 3x + 5$ shown above, $f(-1) = 2$, because $(-1,2)$ is a point on the line.

Features of a function graph:

When looking at the behavior of a function on a graph, it is helpful to identify key features of the graph. We should recognize where the function is positive or negative, increasing or decreasing, and its end behavior. We should also recognize its extrema.

For example: For the discussion to follow, we will refer to the graph below.

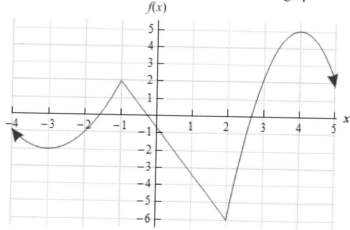

A function is **positive** where its graph lies above the *x*-axis, and **negative** where its graph lies below the *x*-axis. It is **increasing** where the graph goes up, when moving from left to right, and **decreasing** where it goes down. Its **end behavior** describes the function at the arrowheads; that is, at the leftmost or rightmost extremes of the graph.
For example:

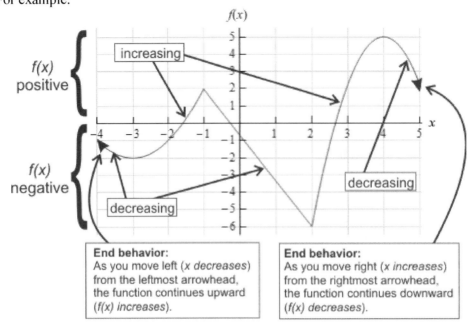

End behavior:
As you move left (*x decreases*) from the leftmost arrowhead, the function continues upward (*f(x) increases*).

End behavior:
As you move right (*x increases*) from the rightmost arrowhead, the function continues downward (*f(x) decreases*).

Some points on the graph of a function can be described as a relative maximum (plural, *maxima*) or a relative minimum (plural, *minima*). A **relative maximum** is a point where no other nearby points have a greater function value (*y*-coordinate), and a **relative minimum** is a point where no other nearby points have a lesser function value. These points are also called **extrema**.
For example:

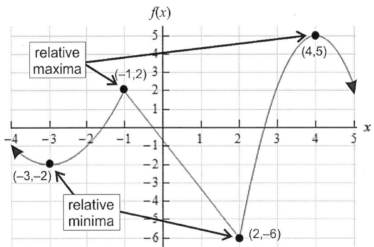

REGENTS QUESTIONS

Multiple Choice

1. A ball is thrown into the air from the edge of a 48-foot-high cliff so that it eventually lands on the ground. The graph below shows the height, *y*, of the ball from the ground after *x* seconds.

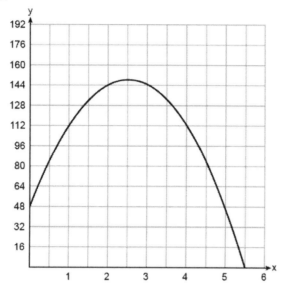

For which interval is the ball's height always *decreasing*?
    (1) $0 \leq x \leq 2.5$ (3) $2.5 < x < 5.5$
    (2) $0 < x < 5.5$ (4) $x \geq 2$

2. The graph of $y = f(x)$ is shown below.

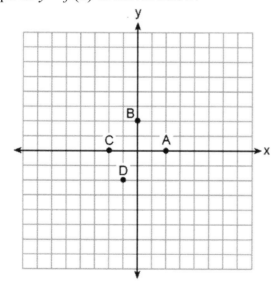

Which point could be used to find $f(2)$?
    (1) A   (3) C
    (2) B   (4) D

Evaluating Functions

Key Terms and Concepts

We saw in the previous section that we can find the value of $f(x)$ for a given x by looking at a graph of the function. More frequently, however, we will use the function's definition. We can find $f(x)$ for a given value of x by **substituting and evaluating** the expression.

For example: For the function $f(x) = 3x + 5$ shown below, $f(-1) = 2$,

(a) because $(-1, 2)$ is a point on the line, and also

(b) because $f(x) = 3x + 5 = 3(-1) + 5 = 2$.

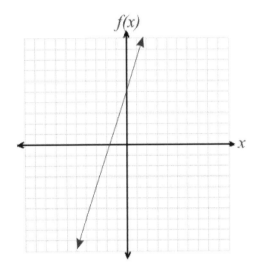

We can also **evaluate a function at a given expression** by replacing the variable in the original function with the expression, using parentheses to avoid errors.

For example: If $f(x) = 3x + 5$, find $f(n+1)$.

$$f(n+1) = 3(n+1) + 5$$
$$= 3n + 3 + 5$$
$$= 3n + 8$$

Practice Problems

1. Find $f(3)$ given $f(x) = -2x^2 - 3x - 6$.	2. $f(m) = 0.5^m$. Evaluate the function for $m = 2$.
3. If $g(x) = 2x^2 + 6x - 3$, find $g(4a)$ in terms of a.	4. If $h(x) = 2x - 1$, find the product $h(0) \cdot h(-2)$.
5. Find $f(a+2)$ in terms of a, given $f(x) = x^2 + 2x - 1$.	6. For what integer value of x is $f(x) = -10$ if $f(x) = -4x + 2$?

REGENTS QUESTIONS

Multiple Choice

1. If $f(x) = kx^2$, and $f(2) = 12$, then k equals
    (1) 1   (3) 3
    (2) 2   (4) 4

2. If $f(x) = |x^3 - 3|$, then $f(-1)$ is equivalent to
    (1) 0   (3) –2
    (2) 2   (4) 4

3. Given: the function f defined by $f(x) = 3x^2 - 4$. Which statement is true?
    (1) $f(0) = 0$ (3) $f(5) + f(2) = f(7)$
    (2) $f(-2) = f(2)$ (4) $f(5) \cdot f(2) = f(10)$

4. The equation $P = 0.0089t^2 + 1.1149t + 78.4491$ models the United States population, P, in millions since 1900. If t represents the number of years after 1900, then what is the estimated population in 2025 to the *nearest tenth of a million*?
    (1) 217.8   (3) 343.9
    (2) 219.0   (4) 356.9

Constructed Response

5. If $f(a) = a^2 - 2a + 1$, find $f(-3)$.

6. If $f(x) = x^2 - 2x + 3$, find $f(-2)$.

7. If $f(x) = (x - 3)^2$, find the value of $f(0)$.

8. If $f(x) = 3x - 4$ and $g(x) = x^2$, find the value of $f(3) - g(2)$.

9. Next weekend Marnie wants to attend either carnival A or carnival B. Carnival A charges $6 for admission and an additional $1.50 per ride. Carnival B charges $2.50 for admission and an additional $2 per ride.
    a) In function notation, write A(x) to represent the total cost of attending carnival A and going on x rides. In function notation, write B(x) to represent the total cost of attending carnival B and going on x rides.
    b) Determine the number of rides Marnie can go on such that the total cost of attending each carnival is the same.
    c) Marnie wants to go on five rides. Determine which carnival would have the lower total cost. Justify your answer.

Operations on Functions

Key Terms and Concepts

We can perform **operations on functions**.

For example: If $f(x) = x + 1$ and $g(x) = x - 2$, then we can find $h(x) = f(x) + g(x)$ by
adding $(x+1) + (x-2) = 2x - 1$, so $h(x) = 2x - 1$.
We can check for any value of x. If $x = 5$, then $f(5) = 5 + 1 = 6$ and
$g(5) = 5 - 2 = 3$, so $f(5) + g(5) = 6 + 3 = 9$. Also, $h(5) = 2(5) - 1 = 9$.

Model Problem

If $f(x) = x^2 + 5$ and $g(x) = 2x - 1$, find $h(x) = f(x) - g(x)$. Find $h(-3)$.

Solution:

$$h(x) = f(x) - g(x)$$

(A) $$= (x^2 + 5) - (2x - 1)$$

(B) $$= x^2 + 5 - 2x + 1$$

$$h(x) = x^2 - 2x + 6$$

(C) $$h(-3) = (-3)^2 - 2(-3) + 6$$
$$= 9 + 6 + 6 = 21$$

Explanation of steps:

(A) Substitute the rule for each function, using parentheses around each expression.
(B) Simplify and express the resulting function rule in standard form.
(C) To evaluate a function for a given value, substitute the value for the variable.
[To check, $h(-3)$ should equal $f(-3) - g(-3)$,
or $\left[(-3)^2 + 5\right] - \left[2(-3) - 1\right] = 14 - (-7) = 21.]$

Practice Problems

1. If $f(x) = x^2 + x + 1$ and $g(x) = x - 5$, find $h(x) = f(x) + g(x)$.	2. If $f(x) = 2x + 1$ and $g(x) = x - 2$, find $h(x) = f(x) \cdot g(x)$.

3. To raise funds, a club is publishing and selling a calendar. The club has sold $500 in advertising and will sell copies of the calendar for $20 each. The cost of printing each calendar is $6. Let c be the number of calendars to be printed and sold.
(a) Write a rule for the function $R(c)$, which gives the revenue generated.
(b) Write a rule for the function $E(c)$, which gives the printing expenses.
(c) Describe how the function $P(c)$, which gives the club's profit, is related to $R(c)$ and $E(c)$, and write a rule for $P(c)$.

Rate of Change for Linear Functions

Key Terms and Concepts

The graph of a function can represent a real event. The **independent** variable is represented by **x-values** in a horizontal axis and the **dependent** variable is represented by **y-values** in a vertical axis. Very often in a real event, time is the independent variable.

A **rate of change** is a rate that describes how one quantity changes in relation to another quantity. If a graph of a function forms a straight line, it is called a **linear** function, and the rate of change can be calculated as the **slope of the line**. Like we saw with *correlation*, if the line has a **positive slope**, there is a **positive rate of change**. If the line has a **negative slope**, there is a **negative rate of change**. If the line is horizontal, the slope is zero and therefore the rate of change is zero.

For example: Every two hours, a driver records the total time and distance traveled. The bivariate table and graph below show the results. A positive slope indicates a *positive rate of change*. Since the slope is 40, the rate is 40 miles per hour.

Time Driving (h)	Distance Traveled (mi)
x	y
2	80
4	160
6	240

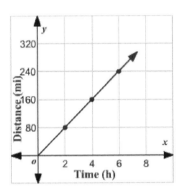

If the line *passes through the origin*, then $y = mx$ and the variables are in **direct variation**. The slope m represents the **constant of variation** when comparing y to x. The above graph is an example of direct variation. The slope, 40, is the constant of variation of *miles* to *hours*.

If the line does not pass through the origin, then the constant term b (the y-intercept) in the equation $y = mx + b$ often represents the **starting value** in the model, especially where the x axis represents time passed.

Model Problem

A candle has a starting length of 10 inches. Thirty minutes after lighting it, the length is 7 inches. The candle continues to get shorter at the same rate over time, as shown by the graph below. Is there a positive or negative rate of change in the length of the candle over time?

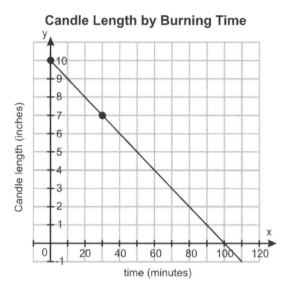

Candle Length by Burning Time

Solution:

Negative

Explanation of Steps:

If the graph shows a line with a positive slope, the rate of change is positive. But if the slope of the line is negative, the rate of change is negative. *[Note that the y-intercept of 10 in the graph represents the starting length of the candle.]*

Practice Problems

1. Identify the rate of change in the following graph as positive, negative or zero.	2. The following table shows a constant rate of change in distance over time. Is the rate of change positive or negative? Calculate the rate.

1. Identify the rate of change in the following graph as positive, negative or zero.

Weight Changes over the Last Six Months

175
150
125
100
75
50
25

Dec Jan Feb Mar Apr May June

2. The following table shows a constant rate of change in distance over time. Is the rate of change positive or negative? Calculate the rate.

Time (hours)	Distance (miles)
4	232
6	348
8	464
10	580

REGENTS QUESTIONS

Multiple Choice

1. If *x* varies directly with *y*, then when *x* is
   (1) multiplied by 2, *y* is multiplied by 2
   (2) multiplied by 2, *y* is divided by 2
   (3) increased by 2, *y* is increased by 2
   (4) increased by 2, *y* is decreased by 2

2. The gas tank in a car holds a total of 16 gallons of gas. The car travels 75 miles on 4 gallons of gas. If the gas tank is full at the beginning of a trip, which graph represents the rate of change in the amount of gas in the tank?

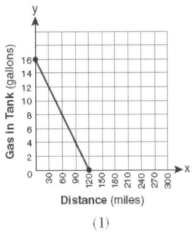

(1)

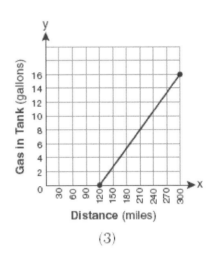

(3)

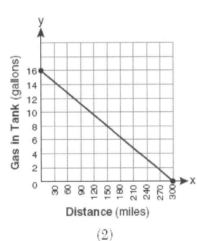

(2)

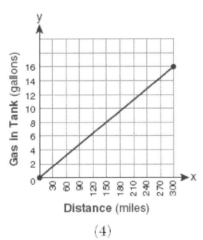

(4)

3. In a linear equation, the independent variable increases at a constant rate while the dependent variable decreases at a constant rate. The slope of this line is
   (1) zero   (3) positive
   (2) negative   (4) undefined

4. The data in the table below are graphed, and the slope is examined.

x	y
0.5	9.0
1	8.75
1.5	8.5
2	8.25
2.5	8.0

The rate of change represented in this table can be described as
(1) negative
(2) positive
(3) undefined
(4) zero

5. In a given linear equation, the value of the independent variable decreases at a constant rate while the value of the dependent variable increases at a constant rate. The slope of this line is
(1) positive
(2) negative
(3) zero
(4) undefined

Average Rate of Change

Key Terms and Concepts

When we calculated the rate of change for linear functions, we simply calculated the slope of the line. For a linear function, the rate of change is constant because the slope is constant. But not all functions are linear. Nevertheless, we can still find an **average rate of change** between any two points on a curve by calculating the slope of the line through those two points, which is called the **secant line**.

As we know, the formula for the slope of a line through points (x_1, y_1) and (x_2, y_2) is

$m = \dfrac{y_2 - y_1}{x_2 - x_1}$. So, using function notation, we could say the average rate of change R over the

interval $a \le x \le b$ is the slope of the secant line through points $(a, f(a))$ and $(b, f(b))$, which

is $R = \dfrac{f(b) - f(a)}{b - a}$. We will use this formula to calculate average rate of change.

For example: Suppose you take a car trip and record the distance that you travel every hour. The graph below shows the distance s (in miles) as a function of time t (in hours). The graph shows that you have traveled a total of 50 miles after 1 hour, 75 miles after 2 hours, 140 miles after 3 hours, and so on.

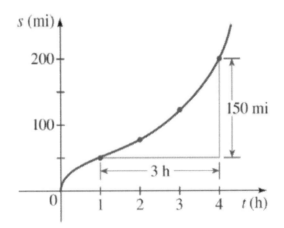

If we want to calculate the average rate of change over the interval $1 \le t \le 4$, we simply find the slope between the points $(1, 50)$ and $(4, 200)$. So, the average rate of change (or average speed in this case) is

$$R = \frac{s(b) - s(a)}{b - a} = \frac{200 - 50}{4 - 1} = \frac{150}{3} = 50 \text{ mph.}$$

For non-linear functions, the **average rate of change may vary** for different intervals.

For example: In the example above, the average rate of change (ie, average speed) over the

interval $2 \le t \le 4$ is $\dfrac{200 - 75}{4 - 2} = \dfrac{125}{2} = 62.5 \text{ mph.}$

274

Model Problem 1

A ball is shot straight up in the air from ground level and its height is recorded every 0.5 seconds until it lands 4 seconds later. A graph of the height of the ball over time (in the shape of a parabola) is shown to the right. Find the average rate of change in the ball's height between seconds 2 and 3.

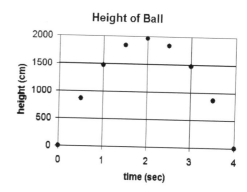

Height of Ball

Solution:

(A) (2,2000) and (3,1500)

(B) $\dfrac{1500-2000}{3-2} = \dfrac{-500}{1} = -500$ cm/sec

Explanation of steps:

(A) Find the points at the start and end of the given interval *[at 2 seconds, the height is at its maximum of 2000cm; at 3 seconds, the height is 1500cm].*

(B) Find the slope of the line between the two points *[a negative slope as it is falling].*

Model Problem 2

If an object is dropped from a tall building, then the distance it has fallen after t seconds is given by the function $d(t) = 16t^2$. Find its average speed (average rate of change) over the interval between 1 second and 5 seconds.

Solution:

(A) At 1 second, $d(1) = 16(1)^2 = 16$. At 5 seconds, $d(5) = 16(5)^2 = 16 \cdot 25 = 400$.

(B) Slope of the line through points (1,16) and (5,400) is $\dfrac{400-16}{5-1} = \dfrac{384}{4} = 96$ ft/sec.

Explanation of steps:

(A) Find the points at the start and end of the given interval. *[Think of each point as the ordered pair of t and d(t). At t = 1, d(t) = 16, and at t = 5, d(t) = 400, so the points are (1,16) and (5,400)]*

(B) Find the slope of the line between the two points.

Practice Problems

1. Calculate the average rate of change of a function, $f(x) = x^2 + 2$ as x changes from 5 to 15?	2. Find the average rate of change for the function $f(x) = x^2 + 10x + 16$ over the interval $-3 \le x \le 3$.

REGENTS QUESTIONS

Multiple Choice

1. Antwaan leaves a cup of hot chocolate on the counter in his kitchen. Which graph is the best representation of the change in temperature of his hot chocolate over time?

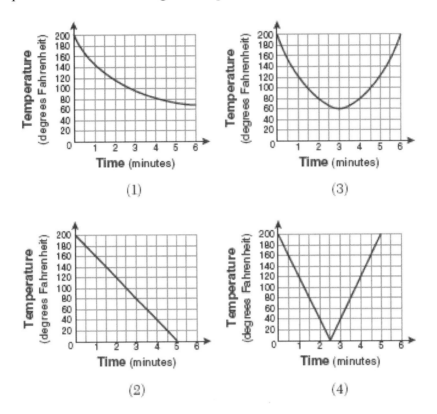

2. Given the functions g(x), f(x), and h(x) shown below:

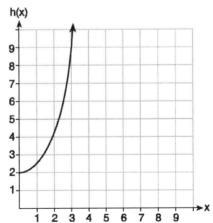

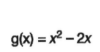

$g(x) = x^2 - 2x$

x	f(x)
0	1
1	2
2	5
3	7

The correct list of functions ordered from greatest to least by average rate of change over the interval $0 \le x \le 3$ is

(1) f(x), g(x), h(x)   (3) g(x), f(x), h(x)

(2) h(x), g(x), f(x)   (4) h(x), f(x), g(x)

3. The Jamison family kept a log of the distance they traveled during a trip, as represented by the graph below.

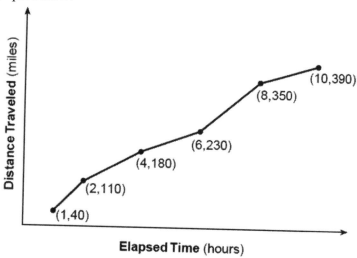

During which interval was their average speed the greatest?
    (1) the first hour to the second hour
    (2) the second hour to the fourth hour
    (3) the sixth hour to the eighth hour
    (4) the eighth hour to the tenth hour

Word Problems – Function Graphs

Key Terms and Concepts

Functions are often used to model real world situations. An important first step in creating graphs of functions to model a situation is to determine what **units** are used for the independent variable's (horizontal) axis and for the dependent variable's (vertical) axis.

For example: To graph a function representing the gasoline in a car's tank during a trip, we could use distance (miles) for one axis and gasoline (gallons) for the other.

The real world situation may also require certain **contraints**, such as minimum or maximum values of the variables. A linear graph, therefore, may be a line segment rather than a line.

For example: The most commonly used units for the independent variables are units of time, which are generally constrained to non-negative real numbers.

Also, there may be **restrictions on the domain**.

For example: If an axis represents a number of people, or a number of items produced, we would restrict its values to counting numbers only.

Since the measurements or values in real world problems are not always small integers, we may need to **scale** our coordinate axes to fit the situation. It is very possible that a grid square in our graph may not represent a one unit by one unit square. We should choose a grid square scale that allows as much of the function's curve to fit on the graph as possible. No matter what scale we choose to use for an axis, we must use consistent intervals on that axis.

For example: A graph below showing world population since 1950 uses a grid square of 5 (in years) by 0.5 (in billions of people).

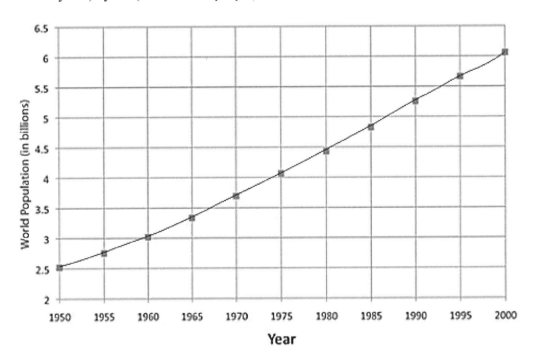

Also, an origin of (0,0) may not always be appropriate. We may choose to use any **origin** as long as the values chosen are less than or equal to as many (if not all) of the data values as possible.

For example: The previous graph starts with the year 1950; it would not be sensible to start at year 0 if we are only interested in the population since 1950. The graph could start with any population of less than or equal to 2.5 billion, the population in 1950, since the population doesn't decrease below that value over time. Therefore, the graph uses (1950, 2) as its origin.

Distance-Time functions: One of the most common types of graphs shows distance over time. Before being able to create such graphs, one needs to know how to interpret them.

If an object moves at a steady speed away from a starting base, the graph will show a straight line with a positive slope (*the distance increases as time increases*). If it returns to base at a steady speed, the graph will show a straight line with a negative slope (*the distance is decreasing over time*). If the object is stationary, we would see a horizontal line with a zero slope (*the distance remains the same as time passes*).

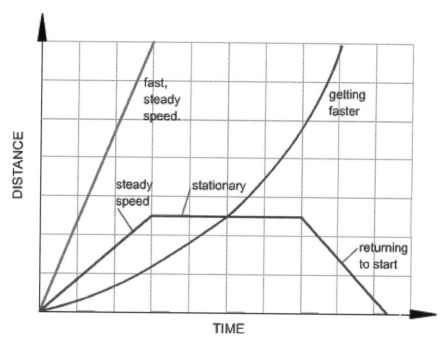

Generally, the horizontal (*x*, or often *t*) axis represents the amount of time passed. A negative value of time would be impossible, so the domain of the function is restricted to non-negative values. The vertical (*y*, or often *d*) axis represents the distance from a starting location. A measure of distance, and thus the range of the function, is also limited to non-negative values.

Model Problem 1: *linear functions*

A swimming pool with a maximum capacity of 450 gallons contains 100 gallons of water before a hose begins to fill the pool by depositing 50 gallons of water each minute. Write and graph an equation that relates x, the number of minutes, to $g(x)$, the number of gallons, for the interval $100 \le g(x) \le 450$ only.

Solution:

(A) $g(x) = 50x + 100$

(B)   (C)

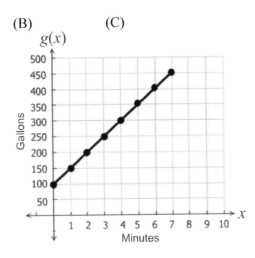

Explanation of steps:

(A) Write an equation.

(B) Create a grid with appropriately scaled axes.

(C) Graph the line. *[Due to the given constraints, the line segment should start at (0,100) and stop when g(x) reaches 450 at (7,450).]*

Model Problem 2: *distance-time graphs*

A car travels away from its starting location at a constant rate of 2 km every 5 minutes for 15 minutes. The car stops and remains idle for the next 20 minutes. It then continues in the same direction at a constant rate of 1 km every 5 minutes for the next 10 minutes. Graph this event on a distance-time graph.

Solution:

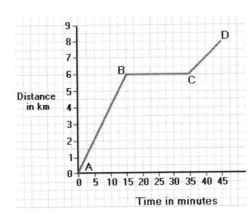

Explanation of steps:

(A) The y-intercept of the graph is the distance at the start time *[we can consider the starting location as a distance of zero]*.

(B) An object moving at a constant rate will show as a straight line. The slope can be determined by moving up (the rise) by a certain distance *[2 km]* and to the right (the run) by a certain amount of time *[5 minutes]*.

(C) A stationary object will not change its distance (y stays the same), but time will pass (x increases), resulting in a horizontal line *[15 + 20 mins = 35 mins]*.

(D) If the distance increases at a slower rate, the line will have a smaller slope *[rise of 1 km and run of 5 minutes]*.

Practice Problems

1. Tom went to the grocery store. The graph below shows Tom's distance from home during his trip. Tom stopped twice to rest on his trip to the store. What is the total amount of time, in minutes, that he spent resting?

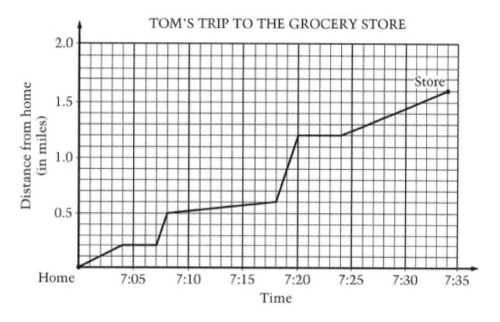

2. Spencer and McKenna are on a long-distance bicycle ride. Spencer leaves one hour before McKenna. The graph below shows each rider's distance in miles from his or her house as a function of time since McKenna left on her bicycle to catch up with Spencer.

    (a) Which function represents Spencer's distance? Which function represents McKenna's distance?

    (b) One rider is speeding up as time passes and the other one is slowing down. Which one is which, and how can you tell from the graphs?

    (c) Estimate when McKenna catches up to Spencer. How far have they traveled at that point in time?

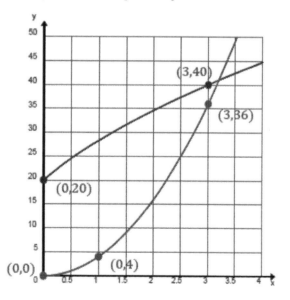

3. The graph below shows distance versus time for a race between runners A and B. The race is already in progress, and the graph shows only the portion of the race that occurred after 11 A.M. The table below lists several characteristics of the graph. Interpret these characteristics in terms of what happened during this portion of the race. Include times and distances to support your interpretation. (*A sample response is given in the table.*)

DISTANCE *VS.* TIME

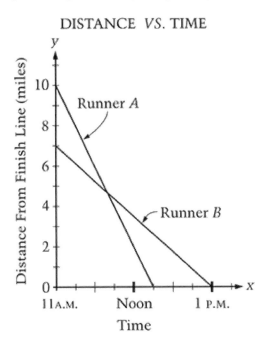

Characteristic of Graph	Interpretation in Terms of the Race
y-intercepts	At 11 A.M. Runner *A* is 10 miles from the finish line and Runner *B* is 7 miles from the finish line.
Slopes	
Point of intersection	
x-intercepts	

REGENTS QUESTIONS

Multiple Choice

1. John left his home and walked 3 blocks to his school, as shown in the accompanying graph.

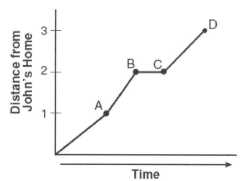

   What is one possible interpretation of the section of the graph from point B to point C?
   (1) John arrived at school and stayed throughout the day.
   (2) John waited before crossing a busy street.
   (3) John returned home to get his mathematics homework.
   (4) John reached the top of a hill and began walking on level ground.

2. The accompanying graph show the amount of water left in Rover's water dish over a period of time.

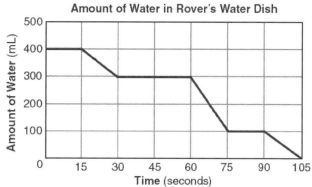

   How long did Rover wait from the end of his first drink to the start of his second drink of water?

   (1) 10 sec   (3) 60 sec
   (2) 30 sec   (4) 75 sec

Constructed Response

3. The accompanying graph shows Marie's distance from home (A) to work (F) at various times during her drive.

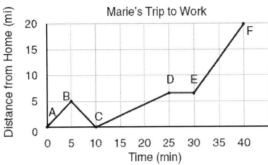

a) Marie left her briefcase at home and had to return to get it. State which point represents when she turned back around to go home and explain how you arrived at that conclusion.

b) Marie also had to wait at the railroad tracks for a train to pass. How long did she wait?

4. During a snowstorm, a meteorologist tracks the amount of accumulating snow. For the first three hours of the storm, the snow fell at a constant rate of one inch per hour. The storm then stopped for two hours and then started again at a constant rate of one-half inch per hour for the next four hours.

a) On the grid below, draw and label a graph that models the accumulation of snow over time using the data the meteorologist collected.

b) If the snowstorm started at 6 p.m., how many inches of snow had accumulated by midnight?

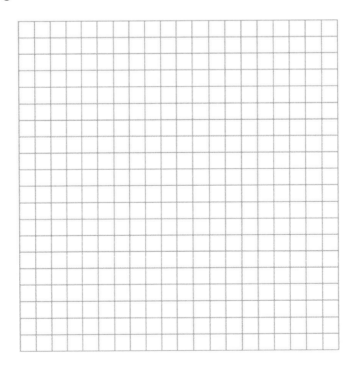

XII.   EXPONENTIAL FUNCTIONS

Percent Change

Key Terms and Concepts

Percent of Change (*Percent Increase* or *Percent Decrease*) =

$$\frac{\text{amount of change}}{\text{original amount}} = \frac{|\text{original} - \text{new}|}{\text{original}} \quad \text{written as a percent}$$

For example: Last month's rent was \$600. This month, it was increased to \$630. What was the percent of increase?

$$\frac{|600 - 630|}{600} = \frac{30}{600} = .05 = 5\% \text{ increase}$$

A **percent of discount** is the percent of decrease on the price of an item that is on sale.

For example: An item originally priced at \$10.50 is sold for \$7.00. This is a $33\frac{1}{3}\%$ discount.

$$\frac{|10.50 - 7.00|}{10.50} = \frac{3.50}{10.50} = 0.\overline{333} = 33\frac{1}{3}\% \text{ discount}$$

Model Problem
The staff at a company went from 40 to 29 employees. What is the percent decrease in staff?

Solution:

$$\begin{array}{cccc} \text{(A)} & \text{(B)} & \text{(C)} & \text{(D)} \end{array}$$
$$\frac{|40 - 29|}{40} = \frac{11}{40} = 0.275 = 27.5\% \text{ decrease.}$$

Explanation of Steps:
- (A) Write a fraction with the amount of change in the numerator and the original amount in the denominator.
- (B) Simplify.
- (C) Divide to change the fraction into a decimal.
- (D) Convert the decimal to a percent by moving the decimal point two places to the right. State whether the change was an increase or decrease.

Practice Problems

1. Zack Cole shot an 86 on the golf course on Monday. On Friday he shot a 95. What is the percent of change?	2. There were 160 candies in a box yesterday, but now there are 116, what is the percent of change?
3. Ann works in a supermarket for $10.00 per hour. If her pay is increased to $12.00, then what is her percent increase in pay?	4. The world population was 4.2 billion people in 1982. The population in 1999 reached 6 billion. Find the percent of change from 1982 to 1999.
5. A quality pen that normally costs $20 is being sold for only $12. Calculate the discount as a percentage.	6. The regular price of a sweater is $40. The sale price is $34. What is the discount as a percentage?

REGENTS QUESTIONS

Multiple Choice

1. A television set which usually sells for $200 is on sale for $170. What is the percent of discount?

    (1) 10%   (3) 18%

    (2) 15%   (4) 30%

Constructed Response

2. The world population was 4.2 billion people in 1982. The population in 1999 reached 6 billion. Find the percent of change from 1982 to 1999, rounded to the *nearest percent*.

3. The Hudson Record Store is having a going-out-of-business sale. CDs normally sell for $18.00. During the first week of the sale, all CDs will sell for $15.00. Written as a fraction, what is the rate of discount? What is this rate expressed as a percent? Round your answer to the *nearest hundredth of a percent*. During the second week of the sale, the same CDs will be on sale for 25% off the *original* price. What is the price of a CD during the second week of the sale?

4. At the end of week one, a stock had increased in value from $5.75 a share to $7.50 a share. Find the percent of increase at the end of week one to the nearest tenth of a percent. At the end of week two, the same stock had decreased in value from $7.50 to $5.75. Is the percent of decrease at the end of week two the same as the percent of increase at the end of week one? Justify your answer.

Exponential Growth and Decay

Key Terms and Concepts

When an amount is **increased** by a certain percentage rate, r, we can calculate the new value by multiplying the original amount by $1 + r$, since $1 = 100\%$.

For example: A \$50 investment increases by 4%. To calculate the new value,
$$50(1 + 0.04) = 50(1.04) = \$52.$$

Similarly, if an amount is **decreased** by a rate of r, we can multiply the amount by $1 - r$ to find the new value, since $1 = 100\%$.

For example: A \$50 investment loses 6% of its value. It is now worth
$$50(1 - 0.06) = 50(0.94) = \$47.$$
                     A 6% loss results in a new amount that is 94% of the original amount.

Often, an increase or decrease is calculated in regular intervals of time. This represents **exponential growth** (increase) or **exponential decay** (decrease).

An example of exponential growth is compound interest. Suppose a \$2,000 investment (principal) is invested in an account which earns 5% interest compounded annually over 4 years.
This is represented by $2000(1.05)(1.05)(1.05)(1.05) = 2000(1.05)^4$.

An example of exponential decay is depreciation. Suppose an industrial machine originally purchased for \$50,000 loses 10% of its value every year for 5 years.
This is represented by $50000(0.90)(0.90)(0.90)(0.90)(0.90) = 50000(0.90)^5$

The **formula for exponential growth** is $y = a(1 + r)^x$, where a is the original amount, r is the constant rate of *increase*, x is the number of times the rate is applied, and y is the final amount.
The **formula for exponential decay** is $y = a(1 - r)^x$ where r is the constant rate of *decrease*.

If we set b to the growth $(1 + r)$ or decay $(1 - r)$ factor, we can build a **general formula for exponential growth or decay**: $y = ab^x$ where $a > 0, b > 0, b \neq 1$. The formula represents *growth* when $b > 1$ or *decay* when $0 < b < 1$. The variable x is usually given in units of time.

Model Problem

The principal of a school predicts that enrollment at her school will increase by 15% each year for the next 4 years. If the current enrollment is 400 students, what would the enrollment be after the fourth year, *to the nearest whole number*, if the principal's predictions are true?

Solution:
    (A)   (B)
$$y = 400(1.15)^4 = 699.6025 \approx 700$$

Explanation of Steps:
    (A) Use the formula $y = ab^x$, where a is the original amount *[400]*, b is the growth factor *[115% or 1.15]*, and x is the number of times the rate is applied *[4, once each year]*.
    (B) Use your calculator to find the result, rounding as directed.

Practice Problems

1. If x is the starting value before an exponential growth of 10% per day, write an expression for the value after 20 days.	2. If x is the starting value before an exponential decay of 2% per day, write an expression for the value after n days.
3. If you deposit $1500 in an account that pays 5% interest yearly, how much money do you have after 6 years?	4. A mouse population is 25,000 and is decreasing at a rate of 20% per year. What is the population after 3 years?
5. The population of Jacksonville is 3,810 and is growing at an annual rate of 3.5%. If this growth rate continues, what will be the approximate population in five years?	6. In a certain game tournament, 75% of the players are eliminated each round. If the tournament starts with 256 players, how many remain after three rounds?
7. A used car was purchased in July 1999 for $11,900. If the car depreciates (loses) 13% of its value each year, what is the value of the car, to the nearest hundred dollars, in July 2002?	8. On January 1, 1999, the price of gasoline was $1.39 per gallon. If the price increased by 0.5% per month, what was the cost of one gallon of gasoline, to the nearest cent, on January 1, 2000?

REGENTS QUESTIONS

Multiple Choice

1. Daniel's Print Shop purchased a new printer for $35,000. Each year it depreciates (loses value) at a rate of 5%. What will its approximate value be at the end of the fourth year?
    - (1) $33,250.00
    - (2) $30,008.13
    - (3) $28,507.72
    - (4) $27,082.33

2. Kathy plans to purchase a car that depreciates (loses value) at a rate of 14% per year. The initial cost of the car is $21,000. Which equation represents the value, *v*, of the car after 3 years?
    - (1) $v = 21{,}000(0.14)^3$
    - (2) $v = 21{,}000(0.86)^3$
    - (3) $v = 21{,}000(1.14)^3$
    - (4) $v = 21{,}000(0.86)(3)$

3. The New York Volleyball Association invited 64 teams to compete in a tournament. After each round, half of the teams were eliminated. Which equation represents the number of teams, *t*, that remained in the tournament after *r* rounds?
    - (1) $t = 64(r)^{0.5}$
    - (2) $t = 64(-0.5)^r$
    - (3) $t = 64(1.5)^r$
    - (4) $t = 64(0.5)^r$

4. Cassandra bought an antique dresser for $500. If the value of her dresser increases 6% annually, what will be the value of Cassandra's dresser at the end of 3 years to the *nearest dollar*?
    - (1) $415
    - (2) $590
    - (3) $596
    - (4) $770

5. In a science fiction novel, the main character found a mysterious rock that decreased in size each day. The table below shows the part of the rock that remained at noon on successive days.

Day	Fractional Part of the Rock Remaining
1	1
2	$\frac{1}{2}$
3	$\frac{1}{4}$
4	$\frac{1}{8}$

Which fractional part of the rock will remain at noon on day 7?
    - (1) $\dfrac{1}{128}$
    - (2) $\dfrac{1}{64}$
    - (3) $\dfrac{1}{14}$
    - (4) $\dfrac{1}{12}$

6. The value, y, of a $15,000 investment over x years is represented by the equation

    $y = 15000(1.2)^{\frac{x}{3}}$. What is the profit (interest) on a 6-year investment?

    (1) $6,600
    (2) $10,799
    (3) $21,600
    (4) $25,799

7. The value of a car purchased for $20,000 decreases at a rate of 12% per year. What will be the value of the car after 3 years?

    (1) $12,800.00
    (2) $13,629.44
    (3) $17,600.00
    (4) $28,098.56

8. The current student population of the Brentwood Student Center is 2,000. The enrollment at the center increases at a rate of 4% each year. To the *nearest whole number*, what will the student population be closest to in 3 years?

    (1) 2,240
    (2) 2,250
    (3) 5,488
    (4) 6,240

9. Mr. Smith invested $2,500 in a savings account that earns 3% interest compounded annually. He made no additional deposits or withdrawals. Which expression can be used to determine the number of dollars in this account at the end of 4 years?

    (1) $2500(1+0.03)^4$
    (2) $2500(1+0.3)^4$
    (3) $2500(1+0.04)^3$
    (4) $2500(1+0.4)^3$

10. A car depreciates (loses value) at a rate of 4.5% annually. Greg purchased a car for $12,500. Which equation can be used to determine the value of the car, V, after 5 years?

    (1) $V = 12,500(0.55)^5$
    (2) $V = 12,500(0.955)^5$
    (3) $V = 12,500(1.045)^5$
    (4) $V = 12,500(1.45)^5$

11. Is the equation $A = 21000(1-0.12)^t$ a model of exponential growth or exponential decay, and what is the rate (percent) of change per time period?

    (1) exponential growth and 12%
    (2) exponential growth and 88%
    (3) exponential decay and 12%
    (4) exponential decay and 88%

12. The current population of a town is 10,000. If the population, P, increases by 20% each year, which equation could be used to find the population after t years?

    (1) $P = 10,000(0.2)^t$
    (2) $P = 10,000(0.8)^t$
    (3) $P = 10,000(1.2)^t$
    (4) $P = 10,000(1.8)^t$

13. Which graph represents the exponential decay of a radioactive element?

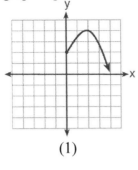

(1)

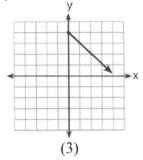

(3)

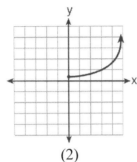

(2)

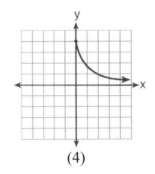

(4)

Constructed Response

14. A bank is advertising that new customers can open a savings account with a $3\frac{3}{4}\%$ interest rate compounded annually. Robert invests $5,000 in an account at this rate. If he makes no additional deposits or withdrawals on his account, find the amount of money he will have, to the *nearest cent*, after three years.

15. The Booster Club raised $30,000 for a sports fund. No more money will be placed into the fund. Each year the fund will decrease by 5%. Determine the amount of money, to the *nearest cent*, that will be left in the sports fund after 4 years.

16. Adrianne invested $2000 in an account at a 3.5% interest rate compounded annually. She made no deposits or withdrawals on the account for 4 years. Determine, to the *nearest dollar*, the balance in the account after the 4 years.

17. Kirsten invested $1000 in an account at an annual interest rate of 3%. She made no deposits or withdrawals on the account for 5 years. The interest was compounded annually. Find the balance in the account, to the *nearest cent*, at the end of 5 years.

18. The breakdown of a sample of a chemical compound is represented by the function $p(t) = 300(0.5)^t$, where $p(t)$ represents the number of milligrams of the substance and t represents the time, in years. In the function $p(t)$, explain what 0.5 and 300 represent.

292

Exponential Functions

Key Terms and Concepts

An **exponential function** is a function in which x appears as an exponent in the equation. For the sake of this course, equations are limited to the form $y = ab^x$ where $a \neq 0$, $b > 0$, and $b \neq 1$.

For examples: $y = 5^x$ or $y = -3(0.5)^x$

An exponential function can be graphed using a table or a graphing calculator.

For example: We can graph $y = 2^x$ as follows.

x	2^x	y	(x,y)
-1	2^{-1}	$\frac{1}{2}$	$(-1,\frac{1}{2})$
0	2^0	1	$(0,1)$
1	2^1	2	$(1,2)$
2	2^2	4	$(2,4)$
3	2^3	8	$(3,8)$

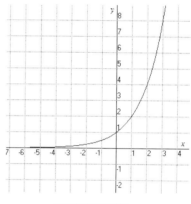

 On the calculator, we would enter: [Y=] 2 [^] [ALPHA] [X] [ZOOM] [ZStandard]

Generally, the graph of an exponential function will be shaped like one of these:

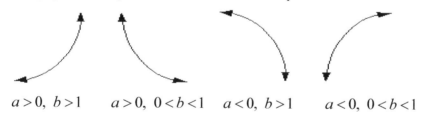

$a > 0,\ b > 1$ $a > 0,\ 0 < b < 1$ $a < 0,\ b > 1$ $a < 0,\ 0 < b < 1$

For example: The graph of the function $y = \left(\frac{1}{2}\right)^x$ looks like this.

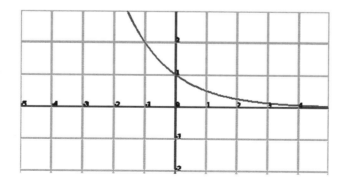

In general, $y = b^x$ will **increase** over the entire domain when $b > 1$ or **decrease** over the entire domain when $0 < b < 1$.

In the function $y = ab^x$, the **constant a** will tell us where the curve crosses the y-axis.

For example: For the function $y = 100\left(\frac{1}{2}\right)^x$, where $a = 100$ and $b = \frac{1}{2}$, the graph would intersect the y-axis at $(0, 100)$, as shown below.

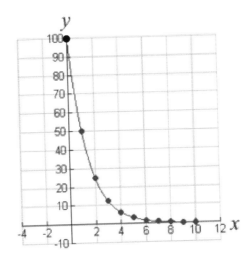

An exponential function $y = ab^x$ will **never actually touch** (intersect) the x-axis. Neither a nor b are equal to zero, so ab^x (and therefore y) will never equal zero.

When **a is negative** in an exponential function, the graph is negative over the entire domain.

For example: The graph of $y = -3^x$ will look like the graph of $y = 3^x$ but **reflected** (flipped) over the x-axis.

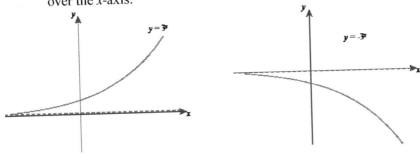

Exponential functions may be used to represent **exponential growth** and **exponential decay**. In these special cases, the equation $y = ab^x$ is limited to $a > 0$.

- When $b > 1$, the function shows **exponential growth**.
- When $0 < b < 1$, the function shows **exponential decay**.

OR

exponential growth   exponential decay

Model Problem

Graph the function $y = \frac{3}{2} \cdot 2^x$.

Solution:

(A) x	(B) $\frac{3}{2} \cdot 2^x$	(C) y	(D) (x,y)
0	$\frac{3}{2} \cdot 2^0 = \frac{3}{2} \cdot 1$	1.5	(0,1.5)
1	$\frac{3}{2} \cdot 2^1 = \frac{3}{2} \cdot 2$	3	(1,3)
2	$\frac{3}{2} \cdot 2^2 = \frac{3}{2} \cdot 4$	6	(2,6)
3	$\frac{3}{2} \cdot 2^3 = \frac{3}{2} \cdot 8$	12	(3,12)
4	$\frac{3}{2} \cdot 2^4 = \frac{3}{2} \cdot 16$	24	(4,24)

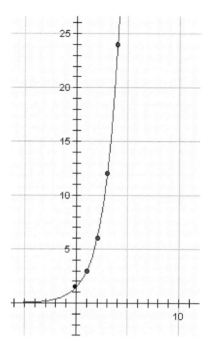

Explanation of Steps:

    (A) Pick values of x.

    (B) Substitute the values of x into the expression.

    (C) Evaluate for y.

    (D) Plot the resulting points on the graph.

Practice Problems

1. On separate graph paper, graph $y = 3^x$.	2. On separate graph paper, graph $y = \left(\frac{1}{3}\right)^x$.
3. On your calculator or on separate graph paper, graph $y = \frac{1}{3} \cdot 2^x$.	4. On your calculator or separate graph paper, graph $y = 3 \cdot \left(\frac{1}{2}\right)^x$.
5. On the calculator, graph $y = 12(1.5)^x$.	6. On the calculator, graph $y = 12(0.5)^x$.
7. On separate graph paper, graph $y = -(2^x)$.	8. On separate graph paper, graph $y = 2^x - 5$ and explain how this graph differs from the exponential function $y = 2^x$.
9. When a piece of paper is folded in half, the total thickness doubles. An unfolded piece of paper is 0.1 millimeter thick. (a) Write an equation for the total thickness, t, as a function of the number of folds, n. (b) Graph the function for a discrete domain of whole numbers, $0 \leq n \leq 8$.	10. The height, h, of a dropped ball is an exponential function of bounces, n. On its first bounce, a certain ball reached a height of 15 inches. On the second bounce, the ball reached a height of 7.5 inches. (a) Write an equation for the height of the ball as a function of the number of bounces. (b) Graph the function for a discrete domain of whole numbers, $0 \leq n \leq 6$.

REGENTS QUESTIONS

Multiple Choice

1. The graph of the equation $y = 2^x$ intersects
   - (1) the *x*-axis, only
   - (2) the *y*-axis, only
   - (3) the *x*-axis and the *y*-axis
   - (4) neither the *x*-axis nor the *y*-axis

2. The graph of the function $y = 3^x$ lies in which quadrant(s)?
   - (1) I, only
   - (2) I and II
   - (3) I and III
   - (4) I and IV

3. On January 1, a share of a certain stock cost $180. Each month thereafter, the cost of a share of this stock decreased by one-third. If *x* represents the time, in months, and *y* represents the cost of the stock, in dollars, which graph best represents the cost of a share over the following 5 months?

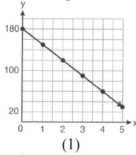

(1)

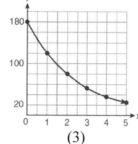

(3)

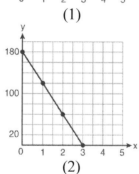

(2)

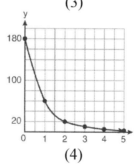

(4)

Constructed Response

4. On the set of axes below, draw the graph of $y = 2^x$ over the interval $-1 \leq x \leq 3$. Will this graph ever intersect the x-axis? Justify your answer.

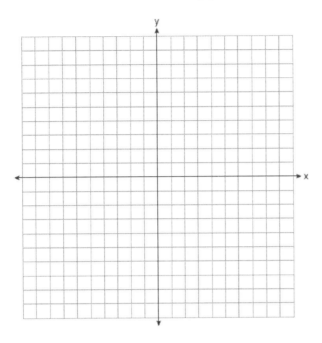

5. On the set of axes below, graph $y = 3^x$ over the interval $-1 \leq x \leq 2$.

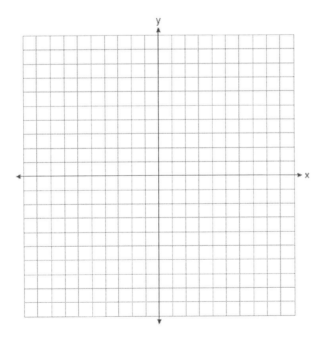

Sequences

Key Terms and Concepts

A **sequence** is an ordered list of numbers. We can often define a sequence as a discrete function. If we let the integer n represent the term number in the sequence, then $f(n)$ represents the nth term in the sequence. Although we may use the set of integers as the domain for a sequence, for this course we will restrict the domain for a sequence to the set of counting or whole numbers.

For example: The sequence of perfect squares is 1, 4, 9, 16, ...

For this sequence, $f(1)=1$, $f(2)=4$, $f(3)=9$, $f(4)=16$, etc.

So, we can define this sequence as the function $f(n)=n^2$.

More commonly, terms of sequences are written using **subscript numbers**; however, this is simply a difference in notation, and either method may be used.

For example: We could have defined $f(n)$ above as $a_n = n^2$. Using this notation, the first four terms would have been written as $a_1 = 1$, $a_2 = 4$, $a_3 = 9$, and $a_4 = 16$.

Arithmetic Sequences

An **arithmetic sequence** is one in which each term is obtained by *adding* the same number (called the **common difference**, represented by d) to the preceding term.

For example: 9, 11, 13, 15, ... $a_1 = 9$, $d = 2$

For an arithmetic sequence, we can define the function using $a_n = a_1 + (n-1)d$ where a_1 is the first term and d is the common difference. By distributing the d, we can simplify the expression. *(This rule is given on the Reference Sheet at the back of the Regents exam.)*

An arithmetic sequence is always a **linear** function.

For example: For the arithmetic sequence 9, 11, 13, 15, ..., $a_1 = 9$ and $d = 2$, so we can define the sequence by substituting for a_1 and d in $a_n = a_1 + (n-1)d$, giving us the function $a_n = 9 + (n-1) \cdot 2$. By simplifying, $a_n = 2n + 7$.

The arithmetic sequence is graphed as a linear function below. Since the function is discrete, the graph consists of isolated points, not a continuous line.

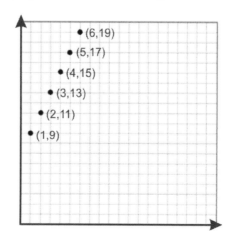

n	$a_n = 2n+7$
1	9
2	11
3	13
4	15
5	17
6	19

As can be seen by the table, we can **add** d to each term to obtain the next term.

Geometric Sequences

A **geometric sequence** is one in which each term is obtained by *multiplying* the same number (called the **common ratio**, represented by r) to the preceding term.

For example: 5, 10, 20, 40, ... $\quad a_1 = 5$, $r = 2$

For a geometric sequence, we can define the function using $a_n = a_1 r^{n-1}$ where a_1 is the first term and r is the common ratio ($r \neq 0$).
(This rule is given on the Reference Sheet at the back of the Regents exam.)

A geometric sequence is always an **exponential** function.

For example: For the geometric sequence 5, 10, 20, 40, ..., $a_1 = 5$ and $r = 2$, so we can

define the sequence by substituting for a_1 and r in $a_n = a_1 r^{n-1}$, giving us

the function $a_n = 5 \cdot 2^{n-1}$.

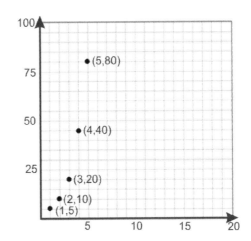

n	$a_n = 5 \cdot 2^{n-1}$
1	5
2	10
3	20
4	40
5	80
6	160

As can be seen by the table, we can **multiply** each term by r to obtain the next term. In this graph, the set of points that make up the discrete function lie on an exponential curve.

Recursive Definitions

Until now, we defined sequences using **explicit definitions** as functions in terms of n. Sequences can often be defined using **recursive definitions**. A recursive definition first defines the starting term (called the *seed term*), and then defines the next terms based on values of previous terms. We can define $f(n)$ in terms of $f(n-1)$ or define $f(n+1)$ in terms of $f(n)$.

For examples: (a) The arithmetic sequence 9, 11, 13, 15, ..., could have been defined by stating that $f(1) = 9$ and $f(n) = f(n-1) + 2$ for $n > 1$. This means the nth term $f(n)$ is defined as the value of the previous term $f(n-1)$ plus 2.

$$\begin{cases} f(1) = 9 \\ f(n) = f(n-1) + 2 \text{ for } n > 1 \end{cases}$$

Alternately, we could have replaced the second part of this definition with $f(n+1) = f(n) + 2$ for $n \geq 1$.

(b) Similarly, the geometric sequence 5, 10, 20, 40, ..., could have been defined by stating that $f(1) = 5$ and $f(n) = 2 \cdot f(n-1)$ for $n > 1$. Or, we could write this as $f(1) = 5$ and $f(n+1) = 2 \cdot f(n)$ for $n \geq 1$.

300

A famous recursively defined sequence is the **Fibonacci sequence**, in which each term (after the first two seed terms) equals the sum of the two previous terms: 1, 1, 2, 3, 5, 8, 13, 21, 34, ... We can define this as $f(1) = 1$, $f(2) = 1$, and $f(n) = f(n-1) + f(n-2)$ for $n > 2$.

Model Problem

Find the 15th term of the arithmetic sequence 5, 11, 17, 23, ...

Solution:	**Explanation of steps:**
(A) $a_1 = 5$, $d = 6$	(A) Find a_1 and d.
	[a_1 is the first term, d is the common difference.]
(B) $a_n = a_1 + (n-1)d$	(B) Write the function rule for arithmetic sequences.
(C) $a_{15} = 5 + (15-1) \cdot 6$	(C) Substitute for a_1, d, and the term number, n.
(D) $a_{15} = 89$	(D) Evaluate.

Practice Problems

1. What is the common difference in the sequence 8, 4, 0, –4, ...?	2. What is the common ratio in the sequence 12, 6, 3, 1.5, ...?
3. Write a function for the *n*th term of the arithmetic sequence, 15, 20, 25, 30, ...	4. Write a function for the *n*th term of the geometric sequence, –1, 2, –4, 8, ...
5. Find the eighth term of the arithmetic sequence for which $a_1 = 21$ and $d = 9$.	6. Find the seventh term of the geometric sequence for which $a_1 = 6$ and $r = -\dfrac{1}{2}$.
7. A sequence is recursively defined as: $a_1 = 3$ and $a_n = a_{n-1} + n$ Write the first four terms of this sequence. Is this sequence arithmetic, geometric, or neither? Justify your answer.	8. The first four terms in a sequence are 40, 8, 24, 16, ... Each term after the first two terms is found by taking one-half the sum of the two preceding terms. (a) Write a recursive definition for this sequence. (b) Which term is the first odd number in this sequence?

REGENTS QUESTIONS

Multiple Choice

1. The population growth of Boomtown is shown in the accompanying graph.

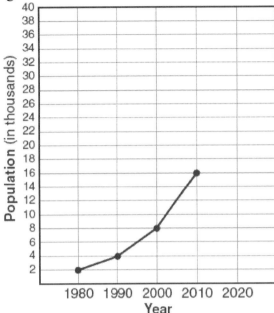

If the same pattern of population growth continues, what will the population of Boomtown be in the year 2020?

   (1) 20,000   (3) 40,000

   (2) 32,000   (4) 64,000

2. What is the common difference of the arithmetic sequence 5, 8, 11, 14, …?

   (1) $\dfrac{8}{5}$ (3) 3

   (2) –3   (4) 9

3. What is a formula for the *n*th term of sequence *B* shown below?

          $B = 10, 12, 14, 16, …$

   (1) $b_n = 8 + 2n$ (3) $b_n = 10(2)^n$

   (2) $b_n = 10 + 2n$ (4) $b_n = 10(2)^{n-1}$

4. What is the fifteenth term of the sequence 5, –10, 20, –40, 80, …?

     (1) –163,840   (3) 81,920

     (2) –81,920   (4) 327,680

5. A sequence has the following terms: $a_1 = 4$, $a_2 = 10$, $a_3 = 25$, $a_4 = 62.5$. Which formula represents the *n*th term in the sequence?

    (1) $a_n = 4 + 2.5n$ (3) $a_n = 4(2.5)^n$

    (2) $a_n = 4 + 2.5(n-1)$ (4) $a_n = 4(2.5)^{n-1}$

6.   What is the common ratio of the geometric sequence shown below?
          $-2, 4, -8, 16, \ldots$

  (1) $-\dfrac{1}{2}$ (3) -2

  (2) 2 (4) -6

7.   A sunflower is 3 inches tall at week 0 and grows 2 inches each week. Which function(s) shown below can be used to determine the height, $f(n)$, of the sunflower in n weeks?

  I. $f(n) = 2n + 3$
  II. $f(n) = 2n + 3(n-1)$
  III. $f(n) = f(n-1) + 2$ where $f(0) = 3$

  (1) I and II   (3) III, only
  (2) II, only   (4) I and III

8.   The diagrams below represent the first three terms of a sequence.

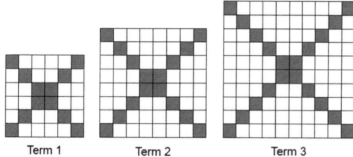

  Term 1   Term 2   Term 3

  Assuming the pattern continues, which formula determines a_n, the number of shaded squares in the nth term?

  (1) $a_n = 4n + 12$ (3) $a_n = 4n + 4$

  (2) $a_n = 4n + 8$ (4) $a_n = 4n + 2$

Constructed Response

9.   Find the first four terms of the recursive sequence defined below.
          $a_1 = -3$

          $a_n = a_{(n-1)} - n$

10. Find the third term in the recursive sequence $a_{k+1} = 2a_k - 1$ where $a_1 = 3$.

Comparing Linear and Exponential Functions

<u>Key Terms and Concepts</u>

Using our knowledge about sequences, we can predict whether a continuous function is *linear* or *exponential* before we even graph the function. First, create a table using *equally spaced* values of the domain. Then look at the values of the function. If we can **add** a constant (a *common difference*) to each value to get the next value, the function is **linear**. If we can **multiply** each value by a constant (a *common ratio*) to get the next value, the function is **exponential**.

For example: Compare the tables of the two functions, $f(n) = 2n$ and $f(n) = 2^n$.

We will use values of *n* that are evenly spaced at 1 unit apart, from –2 to 3.

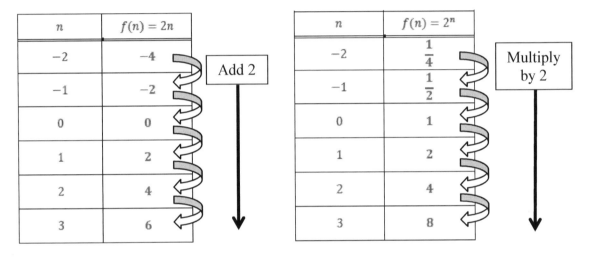

In the first table, we can add a constant, 2, to get the next function value, so this function is linear. In the second table, we can multiply each function value by 2 to get the next value, so the function is exponential.

The graph of the two functions, below, confirms our findings by their shapes:

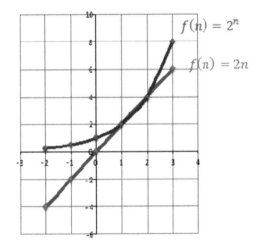

An exponential growth function $g(x)$ will *always eventually exceed* any linear function $f(x)$.

That is, *x* will eventually reach a value *k* for which $g(x) > f(x)$ for all $x > k$.

Practice Problems

1. Identify each function as linear, exponential, or neither.

    a)

x	-3	-2	-1	0	1	2	3
y	14	10	6	2	-2	-6	-10

    c)

x	-3	-2	-1	0	1	2	3
y	$\frac{1}{2}$	1	2	4	8	16	32

    b)

x	-3	-2	-1	0	1	2	3
y	$\frac{1}{9}$	$\frac{1}{3}$	1	3	9	27	81

    d)

x	-3	-2	-1	0	1	2	3
y	-27	-9	-3	0	3	9	27

2. A tile pattern is shown below. Create an explicit formula that could be used to determine the number of squares in the nth figure.

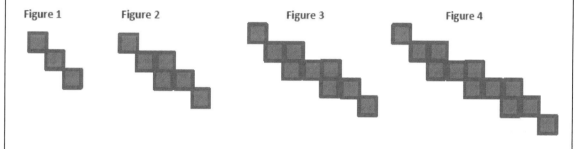

Figure 1   Figure 2   Figure 3   Figure 4

3. A tile pattern is shown below. Create an explicit formula that could be used to determine the number of black triangles in the nth figure.

Figure 1   Figure 2   Figure 3   Figure 4

REGENTS QUESTIONS

Multiple Choice

1. The table below shows the average yearly balance in a savings account where interest is compounded annually. No money is deposited or withdrawn after the initial amount is deposited.

Year	Balance, in Dollars
0	380.00
10	562.49
20	832.63
30	1232.49
40	1824.39
50	2700.54

Which type of function best models the given data?
(1) linear function with a negative rate of change
(2) linear function with a positive rate of change
(3) exponential decay function
(4) exponential growth function

2. The table below represents the function F.

x	3	4	6	7	8
F(x)	9	17	65	129	257

The equation that represents this function is
(1) $F(x) = 3^x$ (3) $F(x) = 2^x + 1$
(2) $F(x) = 3x$ (4) $F(x) = 2x + 3$

Constructed Response

3. Caitlin has a movie rental card worth $175. After she rents the first movie, the card's value is $172.25. After she rents the second movie, its value is $169.50. After she rents the third movie, the card is worth $166.75.

Assuming the pattern continues, write an equation to define $A(n)$, the amount of money on the rental card after n rentals.

Caitlin rents a movie every Friday night. How many weeks in a row can she afford to rent a movie, using her rental card only? Explain how you arrived at your answer.

XIII. FACTORING

Factoring Out the Greatest Common Factor

Key Terms and Concepts

The **distributive property** allows us to express the product of a monomial and polynomial as a single polynomial:

For example: $3x(x+2) = 3x^2 + 6x$

There are times when we would like to factor a polynomial. To **factor** is to break a polynomial down into a product of its factors. Factoring is a process that reverses multiplying.

As part of this process, we may need to find the **greatest common factor (GCF)** of the terms. Remember that we found the GCF of whole numbers by listing their prime factorizations and then determining which factors they had in common.

For example: GCF of 60 and 75. $60 = 2 \cdot 2 \cdot \underline{3} \cdot \underline{5}$ and $75 = \underline{3} \cdot \underline{5} \cdot 5$.

They have $\underline{3} \cdot \underline{5} = 15$ in common, so 15 is the GCF.

If the terms of a polynomial have any factors in common, we can use the distributive property (in reverse) to break the polynomial down into factors.

To factor out the GCF:
  1. Identifying the greatest common factor (GCF) of its terms.
  2. Find the other factor by either
      a. gathering up the remaining factors from each term, or
      b. dividing the original polynomial by the GCF
  3. Write the result as a product of the two factors.

For example: $3x^2 + 6x$ can be rewritten as $\underline{3} \cdot \underline{x} \cdot x + 2 \cdot \underline{3} \cdot \underline{x}$, giving us a GCF of $3x$.

To find the other factor, we can either:

(a) gather up the remaining (*not underlined*) factors from the prime factorization

$\underline{3} \cdot \underline{x} \cdot x + 2 \cdot \underline{3} \cdot \underline{x}$ to get $x+2$, or

(b) we can divide the original polynomial by the GCF, as in $\dfrac{3x^2 + 6x}{3x} = x+2$.

Write the result as the product of the two factors: $3x(x+2)$.

Note: You can check your answer by applying the distributive property.

If all of the factors in a term's prime factorization are underlined as part of the GCF, you are left with a "hidden" factor of 1. So, you'll need to write the term 1 as part of the other factor. Remember that the other factor should have as many terms as the original polynomial.

For example: $3x^2 + 3x$ can be rewritten as $\underline{3} \cdot \underline{x} \cdot x + \underline{3} \cdot \underline{x}$, giving us a GCF of $3x$. When gathering up the remaining factors, we see that the all of the second term's factors are underlined, leaving us with just a "hidden" factor of 1. So, our other factor is $x+1$. The result in factored form is $3x(x+1)$.

The need to use a term of 1 is more easily seen when we divide by the GCF:

$$\frac{3x^2 + 3x}{3x} = \frac{3x^2}{3x} + \frac{3x}{3x} = x+1$$

Model Problem

Factor $8x^2y - 12xy + 20y^2$.

Solution:

(A) $8x^2y - 12xy + 20y^2 = \underline{2} \cdot \underline{2} \cdot 2 \cdot x \cdot x \cdot \underline{y} - \underline{2} \cdot \underline{2} \cdot 3 \cdot x \cdot \underline{y} + \underline{2} \cdot \underline{2} \cdot 5 \cdot \underline{y} \cdot y$

(B) $\qquad\qquad\qquad = 2 \cdot 2 \cdot y(2 \cdot x \cdot x - 3 \cdot x + 5 \cdot y)$

(C) $\qquad\qquad\qquad = 4y(2x^2 - 3x + 5y)$

Explanation of Steps:

(A) Find the GCF of the terms. One way to do this is to expand each term using the prime factorization methods. Then underline any factors that are common to <u>all terms</u>.

(B) The underlined factors represent the GCF $[2 \cdot 2 \cdot y = 4y]$.

The factors that remain (not underlined) should be written in parentheses as the other factor. (If all of a term's factors are underlined, write 1.)

You can also find the second factor by dividing the original polynomial by the GCF

$[\dfrac{8x^2y - 12xy + 20y^2}{4y} = 2x^2 - 3x + 5y]$.

(C) Write the result in unexpanded form.

Practice Problems

1. Factor: $4x^2 - 6x$	2. Factor: $5a^2 - 10a$
3. Factor: $14x^3 + 7x$	4. Factor: $x^3 + x^2 - x$
5. Factor: $12x^3y + 18xy^2$	6. Factor: $2y^3 - 4y^2 + 2y$
7. Factor: $3x^3 - 6x^2 + 6x$	8. Factor: $-2x - 2y$

REGENTS QUESTIONS

Multiple Choice

1. If $3x$ is one factor of $3x^2 - 9x$, what is the other factor?

   (1) $3x$ (3) $x - 3$

   (2) $x^2 - 6x$ (4) $x + 3$

2. The greatest common factor of $3m^2n + 12mn^2$ is

   (1) $3n$ (3) $3mn$

   (2) $3m$ (4) $3mn^2$

Factoring a Trinomial

Key Terms and Concepts

When we multiply two binomials (by FOIL), the result is often a trinomial.

For example: $(x-3)(x+2) = x^2 + 2x - 3x - 6 = x^2 - x - 6$

Therefore, we can often factor a trinomial into the product of two binomial factors.

If the trinomial is a second-degree polynomial in one variable in standard form with a lead coefficient of 1 (that is, if x is the variable, then the polynomial is of the form $x^2 + bx + c$, where b and c are integers), then we can use the following method to factor the trinomial.

To factor $x^2 + bx + c$ by the product-sum method:
(A) Find two integers (if any) that *multiply* to give us c and *add* to give us b.
(B) Factor the trinomial into two binomials in which the variable is written as each first term and the two integers are written as the last terms.

Remember that if c is positive, the two integers must have the same signs, but if c is negative, the two integers must have different signs.

For example: To factor $x^2 - x - 6$ *[b = –1 and c = –6]*, we need two integers whose product is –6 and whose sum is –1. Those two integers are –3 and 2.
So, write the result as $(x-3)(x+2)$.
Note: You can check your answer by multiplying the two binomials.

Note that not all trinomials can be factored. For example, $x^2 + x + 1$ cannot be factored because there are no two integers whose product is 1 and whose sum is 1. A trinomial that cannot be factored is called a **prime trinomial**.

Model Problem

Factor $x^2 - 8x + 12$.

Solution:

$(x-6)(x-2)$

Explanation of Steps:

Find two integers that multiply to give us c [+12] and add to give us b [–8].
[The factors of 12 are: 12 × 1, 6 × 2, and 4 × 3, as well as –12 × –1, –6 × –2, and –4 × –3. The only pair that adds to give us –8 is –6 and –2. So, the answer is $(x-6)(x-2)$.]

Practice Problems

1. Factor $x^2 + 9x + 14$	2. Factor $x^2 - 11x + 18$
3. Factor $x^2 - 6x - 27$	4. Factor $a^2 - a + 210$
5. Determine whether the following trinomial is prime: $x^2 - 3x + 15$	6. Add the trinomials and then factor the result: $(-3x^2 + x - 2) + (4x^2 + 3x - 10)$

REGENTS QUESTIONS

Multiple Choice

1. What are the factors of $x^2 - 10x - 24$?
    - (1) $(x-4)(x+6)$
    - (2) $(x-4)(x-6)$
    - (3) $(x-12)(x+2)$
    - (4) $(x+12)(x-2)$

2. What are the factors of $x^2 - 5x + 6$?
    - (1) $(x+2)$ and $(x+3)$
    - (2) $(x-2)$ and $(x-3)$
    - (3) $(x+6)$ and $(x-1)$
    - (4) $(x-6)$ and $(x+1)$

3. What are the factors of the expression $x^2 + x - 20$?
    - (1) $(x+5)$ and $(x+4)$
    - (2) $(x+5)$ and $(x-4)$
    - (3) $(x-5)$ and $(x+4)$
    - (4) $(x-5)$ and $(x-4)$

4. Which expression is a factor of $x^2 + 2x - 15$?
    - (1) $(x-3)$
    - (2) $(x+3)$
    - (3) $(x+15)$
    - (4) $(x-5)$

5. Which expression is a factor of $n^2 + 3n - 54$?
    - (1) $n+6$
    - (2) $n^2 + 9$
    - (3) $n-9$
    - (4) $n+9$

6. Which is a factor of $x^2 + 5x - 24$
    - (1) $(x+4)$
    - (2) $(x-4)$
    - (3) $(x+3)$
    - (4) $(x-3)$

Constructed Response

7. If $x+2$ is a factor of $x^2 + bx + 10$, what is the value of b?

Factoring the Difference of Two Perfect Squares

Key Terms and Concepts

When multiplying binomials, you may have noticed an interesting result when the binomials are a sum and difference of the same two terms. The middle terms of the product cancel out, leaving us with just the difference of the squares of the terms in the result.

For example: $(x+3)(x-3) = x^2 - 3x + 3x - 9 = x^2 - 9$

So, if we want to factor a **difference of two perfect squares** into its two binomial factors, we can simply take the **sum and difference of the square roots** of the two terms.

For example: $x^2 - 9$ is a difference of two perfect squares.

By taking the square root of each term (x and 3), we can factor this expression into the product of two binomials: $(x+3)(x-3)$.

In order to use this method, *both terms must be perfect squares* and there must be a *subtraction* sign between them.

Be sure to write the square roots in the same order that their squares appear in the problem. For example, $(x+3)(x-3)$ is not the same as $(3+x)(3-x)$; subtraction is not commutative.

Model Problem

Factor $9x^4 - 25$.

Solution:

$(3x^2 + 5)(3x^2 - 5)$

Explanation of Steps:

If the expression is the difference (subtraction) of two perfect squares, then take the square root of each term $[\sqrt{9x^4} = 3x^2 \text{ and } \sqrt{25} = 5]$.

Write the binomial factors as the sum $[3x^2 + 5]$ and difference $[3x^2 - 5]$ of the square roots, and express the answer as a product of the binomial factors $[(3x^2 + 5)(3x^2 - 5)]$.

Practice Problems

1. Factor $x^2 - 36$	2. Factor $4x^2 - 9$
3. Factor $9 - x^2$	4. Factor $a^2 - 1$
5. Factor $49x^2 - y^2$	6. Factor $4a^2 - 9b^2$
7. Factor $x^2y^2 - 16$	8. Factor $x^{10} - 100$

REGENTS QUESTIONS

Multiple Choice

1. The expression $x^2 - 16$ is equivalent to
   (1) $(x+2)(x-8)$ (3) $(x+4)(x-4)$
   (2) $(x-2)(x+8)$ (4) $(x+8)(x-8)$

2. Factored, the expression $16x^2 - 25y^2$ is equivalent to
   (1) $(4x - 5y)(4x + 5y)$ (3) $(8x - 5y)(8x + 5y)$
   (2) $(4x - 5y)(4x - 5y)$ (4) $(8x - 5y)(8x - 5y)$

3. The expression $9x^2 - 100$ is equivalent to
   (1) $(9x - 10)(x + 10)$ (3) $(3x - 100)(3x - 1)$
   (2) $(3x - 10)(3x + 10)$ (4) $(9x - 100)(x + 1)$

4. Which expression is equivalent to $9x^2 - 16$?
   (1) $(3x+4)(3x-4)$ (3) $(3x+8)(3x-8)$
   (2) $(3x-4)(3x-4)$ (4) $(3x-8)(3x-8)$

5. If Ann correctly factors an expression that is the difference of two perfect squares, her factors could be
   (1) $(2x+y)(x-2y)$ (3) $(x-4)(x-4)$
   (2) $(2x+3y)(2x-3y)$ (4) $(2y-5)(y-5)$

6. Which expression is equivalent to $121 - x^2$?
   (1) $(x-11)(x-11)$ (3) $(11-x)(11+x)$
   (2) $(x+11)(x-11)$ (4) $(11-x)(11-x)$

7. The expression $x^2 - 36y^2$ is equivalent to
   (1) $(x-6y)(x-6y)$ (3) $(x+6y)(x-6y)$
   (2) $(x-18y)(x-18y)$ (4) $(x+18y)(x-18y)$

8. Which expression is equivalent to $64 - x^2$?
   (1) $(8-x)(8-x)$ (3) $(x-8)(x-8)$
   (2) $(8-x)(8+x)$ (4) $(x-8)(x+8)$

9. The expression $9a^2 - 64b^2$ is equivalent to
   (1) $(9a-8b)(a+8b)$ (3) $(3a-8b)(3a+8b)$
   (2) $(9a-8b)(a-8b)$ (4) $(3a-8b)(3a-8b)$

10. The expression $100n^2 - 1$ is equivalent to
    (1) $(10n+1)(10n-1)$ (3) $(50n+1)(50n-1)$
    (2) $(10n-1)(10n-1)$ (4) $(50n-1)(50n-1)$

Factoring Completely

Key Terms and Concepts

Sometimes, more than one method of factoring must be performed in order to completely factor an expression. An expression is **factored completely** if all of the factors are prime.

To factor completely:
    (A) Factor out a greatest common factor, if there is one.
    (B) Factor any trinomials, if possible, or any differences of two perfect squares.
    (C) Repeat these steps for each factor until every factor is prime.

Model Problem

Factor completely: $5x^4 - 5$

Solution:

$$5x^4 - 5 =$$

(A) $5(x^4 - 1) =$

(B) $5(x^2 + 1)(x^2 - 1) =$

(C) $5(x^2 + 1)(x + 1)(x - 1)$

Explanation of Steps:

    (A) Factor out the GCF, if any *[the GCF is 5]*.

    (B) If any factor is a trinomial *[not here]* or a difference of two perfect squares *[$x^4 - 1$]*, then factor it.

    (C) Look at each factor to see if any of them can be factored further. *[The factor $x^2 - 1$ is a difference of two perfect squares and can be factored further.]*

Practice Problems

1. Factor completely: $2y^2 + 12y - 54$	2. Factor completely: $3x^2 + 15x - 42$
3. Factor completely: $3x^2 - 27$	4. Factor completely: $2x^2 - 50$
5. Factor completely: $2a^2 - 10a - 28$	6. Factor completely: $x^3 + 8x^2 + 7x$
7. Factor completely: $2x^8 + 16x^7 + 32x^6$	8. Factor completely: $3ax^2 - 27a$
9. Factor completely: $5x^2y^3 - 180y$	10. Factor completely: $2x^5 - 32x$

REGENTS QUESTIONS

Multiple Choice

1. When $2x^2 - 18$ is factored completely, the result is
   - (1) $2(x-3)(x+3)$
   - (3) $(2x+6)(x-3)$
   - (2) $(2x-6)(x+3)$
   - (4) $2(x^2-9)$

2. Factored completely, the expression $2x^2 + 10x - 12$ is equivalent to
   - (1) $2(x-6)(x+1)$
   - (3) $2(x+2)(x+3)$
   - (2) $2(x+6)(x-1)$
   - (4) $2(x-2)(x-3)$

3. Factored completely, the expression $6x - x^3 - x^2$ is equivalent to
   - (1) $x(x+3)(x-2)$
   - (3) $-x(x-3)(x+2)$
   - (2) $x(x-3)(x+2)$
   - (4) $-x(x+3)(x-2)$

4. Factored completely, the expression $3x^2 - 3x - 18$ is equivalent to
   - (1) $3(x^2-x-6)$
   - (3) $(3x-9)(x+2)$
   - (2) $3(x-3)(x+2)$
   - (4) $(3x+6)(x-3)$

5. When $a^3 - 4a$ is factored completely, the result is
   - (1) $(a-2)(a+2)$
   - (3) $a^2(a-4)$
   - (2) $a(a-2)(a+2)$
   - (4) $a(a-2)^2$

6. What are the factors of the expression $x^2 + x - 20$?
   - (1) $(x+5)$ and $(x+4)$
   - (3) $(x-5)$ and $(x+4)$
   - (2) $(x+5)$ and $(x-4)$
   - (4) $(x-5)$ and $(x-4)$

7. Which expression represents $36x^2 - 100y^6$ factored completely?
   - (1) $2(9x+25y^3)(9x-25y^3)$
   - (3) $(6x+10y^3)(6x-10y^3)$
   - (2) $4(3x+5y^3)(3x-5y^3)$
   - (4) $(18x+50y^3)(18x-50y^3)$

8. Factored completely, the expression $3x^3 - 33x^2 + 90x$ is equivalent to
   - (1) $3x(x^2-33x+90)$
   - (3) $3x(x+5)(x+6)$
   - (2) $3x(x^2-11x+30)$
   - (4) $3x(x-5)(x-6)$

9. When factored completely, the expression $3x^2 - 9x + 6$ is equivalent to
   - (1) $(3x-3)(x-2)$
   - (3) $3(x+1)(x-2)$
   - (2) $(3x+3)(x-2)$
   - (4) $3(x-1)(x-2)$

Constructed Response

10. Factor completely: $x^3 - x^2 - 6x$

11. Factor completely: $3x^2 + 15x - 42$

12. Factor completely: $4x^3 - 36x$

13. Factor completely: $5x^3 - 20x^2 - 60x$

14. Factor the expression $x^4 + 6x^2 - 7$ completely.

Factoring by Grouping

Key Terms and Concepts

Another method of factoring allows us to factor four-term polynomials. It is essentially the reverse of the procedure for multiplying two binomials.

Consider the procedure to multiply $(x^2 - 2)(3x + 1)$ by distribution:

$$(x^2 - 2)(3x + 1)$$

(A) $x^2(3x + 1) - 2(3x + 1)$ We multiply each term of the first binomial by the second factor.

(B) $3x^3 + x^2 - 6x - 2$ The result, unless there are like terms, is a four term polynomial.

To factor a 4-term polynomial by grouping:
1. Group the terms into two pairs, where each pair has a common factor.
2. Factor the GCF from each group.
3. Factor out the common binomial (reverse distribute).
4. If either of the binomial factors can be factored further, do so.

For example: Given: $3x^3 + x^2 - 6x - 2$ (*the result from the example above*),

1. Group into pairs of terms: $\boxed{3x^3 + x^2}\ \boxed{-6x - 2}$

2. Factor the GCF from each group: $\boxed{x^2(3x + 1)}\ \boxed{-2(3x + 1)}$

3. Factor out the common binomial $(3x + 1)$ using reverse distribution:

$$\underline{x^2}(3x + 1)\underline{-2}(3x + 1) = (x^2 - 2)(3x + 1)$$

4. Both of these binomial factors are prime.

Model Problem 1
Factor $x^3 + 5x^2 + 10x + 50$.

Solution:

(A) $\boxed{x^3 + 5x^2}\ \boxed{+10x + 50}$

(B) $x^2(x + 5) + 10(x + 5)$

(C) $(x^2 + 10)(x + 5)$

Explanation of steps:
(A) Group the terms into two pairs, where each pair has a common factor.
(B) Factor the GCF from each group.
(C) Factor out the common binomial (reverse distribute). *[Both of the factors are prime.]*

Model Problem 2
Factor $x^3 + x^2 - x - 1$ completely.

Solution:

(A) $\boxed{x^3 + x^2}\ \boxed{-x - 1}$

(B) $x^2(x + 1) - 1(x + 1)$

(C) $(x^2 - 1)(x + 1)$

(D) $(x - 1)(x + 1)(x + 1)$

Explanation of steps:
(A) Group the terms into two pairs, where each pair has a common factor.
(B) Factor the GCF from each group. *[Notice how factoring out the GCF of −1 helps us.]*
(C) Factor out the common binomial (reverse distribute).
(D) Factor either binomial further. *[$(x^2 - 1)$ is a difference of two perfect squares, factored into $(x - 1)(x + 1)$.]*

REGENTS QUESTIONS

Multiple Choice

1. When factored completely, $x^3 + 3x^2 - 4x - 12$ equals

    (1) $(x+2)(x-2)(x-3)$ (3) $(x^2-4)(x+3)$

    (2) $(x+2)(x-2)(x+3)$ (4) $(x^2-4)(x-3)$

2. When factored completely, the expression $3x^3 - 5x^2 - 48x + 80$ is equivalent to

    (1) $(x^2-16)(3x-5)$ (3) $(x+4)(x-4)(3x-5)$

    (2) $(x^2+16)(3x-5)(3x+5)$ (4) $(x+4)(x-4)(3x-5)(3x-5)$

Factoring Trinomials with Leads Other than 1

Key Terms and Concepts

In an earlier section, we saw a simple method for factoring a trinomial for which the lead coefficient is 1. In other words, we factored trinomials of the form $x^2 + bx + c$ where b and c are non-zero integers. When the lead coefficient is a non-zero integer other than 1 (that is, for trinomials of the form $ax^2 + bx + c$ and $a \neq 1$), that simple method can no longer help us. However, as long as they are not prime, we can still factor these types of trinomials. First, we **expand** them into equivalent polynomials with four terms, and then **factor by grouping**.

To factor a trinomial $ax^2 + bx + c$ where a, b, and c are non-zero integers:
1. Find two integers that multiply to give you ac but add to give you b.
2. Break up the middle term into the sum of two terms with these coefficients.
3. Factor by grouping.

Model Problem
Factor $4x^2 - 5x - 6$.

Solution:

(A) $4x^2 - 5x - 6$

$ac = 4(-6) = -24$ and $b = -5$

$+3$ and -8

(B) $4x^2 \boxed{+3x - 8x} - 6$

(C) $\boxed{4x^2 + 3x}\boxed{-8x - 6}$

$x(4x + 3) - 2(4x + 3)$

$(x - 2)(4x + 3)$

Explanation of steps:

(A) Find two integers that multiply to give you ac but add to give you b.

$[3 \times (-8) = -24$ and $3 + (-8) = -5.]$

(B) Break up the middle term into the sum of two terms with these coefficients.

$[-5x \Rightarrow +3x - 8x]$

(C) Factor by grouping.

Practice Problems

1. Factor: $6x^2 + x - 2$	2. Factor: $12x^2 + 5x - 2$
3. Factor: $12x^2 - 29x + 15$	4. Factor: $6x^2 - 11x + 4$
5. Factor: $15x^2 + 14x - 8$	6. Factor: $-10x^2 - 29x - 10$
7. Factor $4x^2 + 12x + 9$. What is the square root of this trinomial, written as a binomial?	8. Factor $-6 + 11x + 10x^2$
9. Factor completely: $5x^2 - 50x + 120$ *(Hint: first factor out the GCF.)*	10. Factor completely: $12x^3 + 14x^2 - 6x$

REGENTS QUESTIONS

Multiple Choice

1. Factored completely, the expression $12x^4 + 10x^3 - 12x^2$ is equivalent to
   (1) $x^2(4x+6)(3x-2)$ (3) $2x^2(2x-3)(3x+2)$
   (2) $2(2x^2+3x)(3x^2-2x)$ (4) $2x^2(2x+3)(3x-2)$

Constructed Response

2. The product of two factors is $2x^2 + x - 6$. One of the factors is $(x+2)$. What is the other factor?

3. Factor completely: $2x^3 - 11x^2 + 5x$

4. Factor completely: $3t^3 + 5t^2 - 12t$

XIV. QUADRATIC EQUATIONS

Solving Quadratic Equations

Key Terms and Concepts

In a **linear equation**, the highest power for the variable x is 1. It is also called a first-degree equation. If an equation in terms of x and y includes an x^2 term, it is called a **quadratic equation**, or a second-degree equation.

We have seen that tables of *linear* functions show a *common difference* in its values (as long as the x values change by constant amounts) and tables of *exponential* functions show a *common ratio*. To recognize quadratic functions, we will not find a constant difference like we do for linear functions, but we'll find that the differences among their differences (called **second differences**) increase by a constant amount, as shown below.

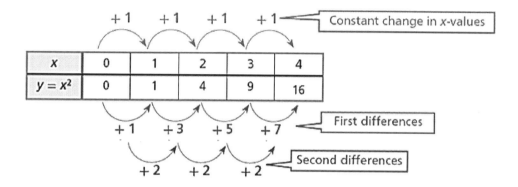

The solutions to a quadratic equation are called its **roots** or **zeros**.

For example: For the equation $x^2 = 25$, the roots (or solutions) are 5 and –5.

The **standard form** for a quadratic equation is $ax^2 + bx + c = 0$, where a, b, and c are integers and zero is on one side of the equation.

To solve a quadratic equation, we use the following fact about a product of two (or more) factors equal to 0. If $ab = 0$, then at least one of the following must be true: $a = 0$ or $b = 0$. This is known as the **zero product property**.

If we can factor a quadratic equation into **factored form**, $g(x - m)(x - n) = 0$ where $g \neq 0$, then the zero product property tells us that either $x - m = 0$ or $x - n = 0$, so $x = m$ or $x = n$. Therefore, the roots are *m* and *n*.

To solve a quadratic equation by factoring:
1. transform the equation into standard form, with zero on one side
2. factor the polynomial completely
3. set each factor that contains a variable equal to zero
4. solve each resulting equation

Model Problem

Find the roots of the equation $x^2 + 6 = 5x$.

Solution:

$$x^2 + 6 = 5x$$

(A) $\dfrac{-5x \quad -5x}{}$

$$x^2 - 5x + 6 = 0$$

(B) $(x-3)(x-2) = 0$

(C) $x - 3 = 0 \quad or \quad x - 2 = 0$

(D) $x = 3 \quad or \quad x = 2$

The roots are 2 and 3.

Explanation of Steps:

(A) transform the equation into standard form, with zero on one side *[by subtracting the 5x from both sides]*.

(B) factor the polynomial completely

(C) set each factor equal to zero

(D) solve each resulting equation

It is also a good idea to check each solution on your calculator. Check the original equation using 2 [STO▸] [ALPHA] [X] then check it again using 3 [STO▸] [ALPHA] [X].

Practice Problems

1. Solve for x: $x^2 - 5x = 0$	2. Solve for x: $x^2 + 3x - 18 = 0$
3. Solve for x: $4x^2 - 36 = 0$	4. Solve for x: $x^2 - 5x - 24 = 0$
5. Solve for x: $x^2 - 5x = 6$	6. Solve for x: $x^2 - 3 = 2x$
7. Solve for x: $x^2 - 4x = x + 24$	8. Solve for x: $2x^2 + 10x = 12$
9. Solve for x: $x(x + 2) = 3$	10. Solve for x: $(x + 2)(x + 3) = 12$

REGENTS QUESTIONS

Multiple Choice

1. What are the roots of the equation $x^2 - 10x + 21 = 0$?
   (1) 1 and 21   (3) 3 and 7
   (2) -5 and -5   (4) -3 and -7

2. What are the roots of the equation $x^2 - 7x + 6 = 0$
   (1) 1 and 7   (3) -1 and -6
   (2) -1 and 7   (4) 1 and 6

3. The solution to the equation $x^2 - 6x = 0$ is
   (1) 0, only   (3) 0 and 6
   (2) 6, only   (4) $\pm\sqrt{6}$

4. What are the roots of the equation $x^2 - 5x + 6 = 0$?
   (1) 1 and –6   (3) –1 and 6
   (2) 2 and 3   (4) –2 and –3

5. The roots of the equation $3x^2 - 27x = 0$ are
   (1) 0 and 9   (3) 0 and 3
   (2) 0 and –9   (4) 0 and –3

6. The roots of the equation $x^2 - 14x + 48 = 0$ are
   (1) –6 and –8   (3) 6 and –8
   (2) –6 and 8   (4) 6 and 8

7. The solutions of $x^2 = 16x - 28$ are
   (1) –2 and –14   (3) –4 and –7
   (2) 2 and 14   (4) 4 and 7

8. The roots of the equation $2x^2 - 8x = 0$ are
   (1) –2 and 2   (3) 0 and –4
   (2) 0, –2 and 2   (4) 0 and 4

9. Find the *positive* root of the equation $3x^2 - 243 = 0$.

10. Solve for x: $x - 2 = \dfrac{3}{x}$

11. Find the roots of the equation $x^2 - x = 6$ algebraically.

12. Find the roots of the equation $x^2 = 30 - 13x$ algebraically.

13. Write an equation that defines $m(x)$ as a trinomial where
$m(x) = (3x - 1)(3 - x) + 4x^2 + 19$.

Solve for x when $m(x) = 0$.

Finding Quadratic Equations from Given Roots

Key Terms and Concepts

If given an equation's roots (zeros), then working backwards we can find the equation.

To find an equation given its roots:
1. Set x equal to each root. For a double root, the same root is used twice.
2. Get zero to one side of each equation.
3. Write as factors.
4. Multiply factors.

Model Problem
Find a quadratic equation with roots of 5 and –1.

Solution:

(A) $x = 5$ or $x = -1$
(B) $x - 5 = 0$ $x + 1 = 0$
(C) $(x-5)(x+1) = 0$
(D) $x^2 - 4x - 5 = 0$

Explanation of steps:
(A) Set x equal to each root.
(B) Get zero to one side of each equation.
(C) Write as factors.
(D) Multiply factors.

Practice Problems

1. Find a quadratic equation with roots of 10 and –2.	2. Find a quadratic equation with roots of 0 and 3.
3. Find a quadratic equation with roots of $\dfrac{3}{2}$ and 2.	4. Find a quadratic equation with the double root of 1.
5. Find a quadratic equation with roots of 4 and –4.	6. When a 3rd degree polynomial is set equal to zero, it is a cubic equation. A cubic equation can have 3 roots. Find a cubic equation with roots of 0,1, and –1.

REGENTS QUESTIONS

Multiple Choice

1. Which quadratic equation has the roots 0 and 3?
    (1) $x^2 = 3$ (3) $x^2 - 3x = 0$
    (2) $x^2 + 3x = 0$ (4) $x^2 - 6x + 9 = 0$

2. For which equation is the solution set $\{-5, 2\}$?
    (1) $x^2 + 3x - 10 = 0$ (3) $x^2 + 3x = -10$
    (2) $x^2 - 3x = 10$ (4) $x^2 - 3x + 10 = 0$

3. Which equation has the solution set $\{1, 3\}$?
    (1) $x^2 - 4x + 3 = 0$ (3) $x^2 + 4x + 3 = 0$
    (2) $x^2 - 4x - 3 = 0$ (4) $x^2 + 4x - 3 = 0$

4. Which equation has roots of -3 and 5?
    (1) $x^2 + 2x - 15 = 0$ (3) $x^2 + 2x + 15 = 0$
    (2) $x^2 - 2x - 15 = 0$ (4) $x^2 - 2x + 15 = 0$

5. If the roots of a quadratic equation are -2 and 3, the equation can be written as
    (1) $(x - 2)(x + 3) = 0$ (3) $(x + 2)(x + 3) = 0$
    (2) $(x + 2)(x - 3) = 0$ (4) $(x - 2)(x - 3) = 0$

6. For which function defined by a polynomial are the zeros of the polynomial -4 and -6?
    (1) $y = x^2 - 10x - 24$ (3) $y = x^2 + 10x - 24$
    (2) $y = x^2 + 10x + 24$ (4) $y = x^2 - 10x + 24$

7. Keith determines the zeros of the function $f(x)$ to be -6 and 5. What could be Keith's function?
    (1) $f(x) = (x + 5)(x + 6)$ (3) $f(x) = (x - 5)(x + 6)$
    (2) $f(x) = (x + 5)(x - 6)$ (4) $f(x) = (x - 5)(x - 6)$

Constructed Response

8. Form the quadratic equation whose roots are -5 and $+7$.

9. The two roots of an equation are -4 and $+3$. Form the equation.

331

Undefined Expressions

Key Terms and Concepts

When we divide $6 \div 3$ we get 2, because when we multiply 2 times 3, the product is 6. However, if we try to divide $6 \div 0$ there is no solution, since there are no values that when multiplied by 0 would give us a product of 6. The expression $6 \div 0$, or $\dfrac{6}{0}$, is undefined.

In fact, any expression involving division is **undefined** when the divisor (denominator) is zero.

Model Problem

For which set of values of x is the algebraic expression $\dfrac{x^2 - 9}{x^2 - 4x}$ undefined?

Solution:

(A) $x^2 - 4x = 0$

(B) $x(x - 4) = 0$

$\{0, 4\}$

Explanation of Steps:

(A) The expression is undefined when the denominator equals zero.

(B) Solve and state the solutions.

Practice Problems

1. For which value of x is the expression $\dfrac{x-7}{x+2}$ undefined?	2. For which value of x is the expression $\dfrac{3x}{3x+1}$ undefined?
3. For which values of x is the expression $\dfrac{2x+3}{x^2-4}$ undefined?	4. For which values of x is the expression $\dfrac{x^2-9}{x^2+2x-8}$ undefined?

REGENTS QUESTIONS

Multiple Choice

1. For which value of x is $\dfrac{x-3}{x^2-4}$ undefined?

    (1) –2   (3) 3

    (2) 0   (4) 4

2. Which value of x makes the expression $\dfrac{x+4}{x-3}$ undefined?

    (1) –4   (3) 3

    (2) –3   (4) 0

3. The function $y=\dfrac{x}{x^2-9}$ is undefined when the value of x is

    (1) 0 or 3   (3) 3, only

    (2) 3 or –3   (4) –3, only

4. Which value of n makes the expression $\dfrac{5n}{2n-1}$ undefined?

    (1) 1   (3) $-\frac{1}{2}$

    (2) 0   (4) $\frac{1}{2}$

5. Which value of x makes the expression $\dfrac{x^2-9}{x^2+7x+10}$ undefined?

    (1) -5   (3) 3

    (2) 2   (4) -3

6. The algebraic expression $\dfrac{x-2}{x^2-9}$ is undefined when x is

    (1) 0   (3) 3

    (2) 2   (4) 9

7. For which set of values of x is the algebraic expression $\dfrac{x^2-16}{x^2-4x-12}$ undefined?

    (1) {–6, 2}   (3) {–4, 4}

    (2) {–4, 3}   (4) {–2, 6}

8. For which values of x is the fraction $\dfrac{x^2+x-6}{x^2+5x-6}$ undefined?

    (1) 1 and –6   (3) 3 and –2

    (2) 2 and –3   (4) 6 and –1

9. The expression $\dfrac{14+x}{x^2-4}$ is undefined when x is

  (1) –14, only   (3) –2 or 2

  (2) 2, only   (4) –14, –2, or 2

10. The expression $\dfrac{x-3}{x+2}$ is undefined when the value of x is

  (1) –2, only   (3) 3, only

  (2) –2 and 3   (4) –3 and 2

11. A value of x that makes the expression $\dfrac{x^2+4x-12}{x^2-2x-15}$ undefined is

  (1) –6   (3) 3

  (2) –2   (4) 5

12. The expression $\dfrac{x-7}{9-x^2}$ is undefined when x is

  (1) 3 and 7   (3) 3, only

  (2) 3 and –3   (4) 9

Solving Proportions by Quadratic Equations

Key Terms and Concepts

Cross-multiplying to solve an algebraic proportion will sometimes result in a quadratic equation.

Model Problem

Solve for x: $\dfrac{4}{x+3} = \dfrac{x-5}{5}$

Solution:

(A) $(x+3)(x-5) = 4 \cdot 5$

(B) $x^2 - 2x - 15 = 20$

(C) $\dfrac{ -20 \ -20}{x^2 - 2x - 35 = 0}$

(D) $(x+5)(x-7) = 0$

(E) $x+5 = 0 \quad or \quad x-7 = 0$

(F) $x = -5 \quad or \quad x = 7$

Solution set is $\{-5, 7\}$

Explanation of Steps:

(A) Cross-multiply.

(B) Simplify both sides.

(C) Transform into standard form by getting a zero to one side.

(D) Factor completely.

(E) Set each variable factor equal to zero.

(F) Solve each resulting equation.

Practice Problems

1. Solve: $\dfrac{1}{x} = \dfrac{x+1}{6}$	2. Solve: $\dfrac{x}{x+3} = \dfrac{5}{x+7}$
3. Solve: $\dfrac{2}{x+1} = \dfrac{x-1}{4}$	4. Solve: $\dfrac{3+x}{2x} = \dfrac{x-1}{x}$

REGENTS QUESTIONS

Multiple Choice

1. Which value of x is a solution of $\dfrac{5}{x} = \dfrac{x+13}{6}$?

    (1) –2   (3) –10
    (2) –3   (4) –15

2. What is the solution set of $\dfrac{x+2}{x-2} = \dfrac{-3}{x}$?

    (1) {–2, 3}   (3) {–1, 6}
    (2) {–3, –2}   (4) {–6, 1}

3. What is the solution of $\dfrac{2}{x+1} = \dfrac{x+1}{2}$?

    (1) –1 and –3   (3) 1 and –3
    (2) –1 and 3   (4) 1 and 3

4. What is the solution of the equation $\dfrac{x+2}{2} = \dfrac{4}{x}$?

    (1) 1 and –8   (3) –1 and 8
    (2) 2 and –4   (4) –2 and 4

Constructed Response

5. Solve for x: $\dfrac{x+1}{x} = \dfrac{-7}{x-12}$

6. Solve algebraically for x: $\dfrac{x+2}{6} = \dfrac{3}{x-1}$

7. Solve algebraically for all values of x: $\dfrac{3}{x+5} = \dfrac{2x}{x^2-8}$

Completing the Square

Key Terms and Concepts

Another method for solving quadratic equations is called **completing the square**. This method converts a quadratic equation into one of the form, $(x + p)^2 = q$, where p and q are constants, allowing the resulting equation to be easily solved by taking the square root of both sides. We do this by manipulating the equation so that the left side is a **perfect square trinomial** that can be factored into the *square of a binomial*, $(x + p)^2$, for some p.

To solve a quadratic equation $ax^2 + bx + c = 0$ **by completing the square:**
1. If a is not a perfect square, multiply (*or divide*) the equation by a. This results in a new equation with new values for a, b, and c.
2. Move the constant term c to the opposite side.
3. Add $\dfrac{b^2}{4a}$ to both sides of the equation.
4. Factor the trinomial into a binomial squared.
5. Take the square root of both sides. Use the $\pm$ symbol on the right side and simplify radicals.
6. Solve for x, remembering that $\pm$ gives two possible solutions.

The advantage of this method is that it can be used when trinomials that cannot be factored. When we use this method on quadratic equations in which *prime trinomials* are set equal to zero, the roots – if there are any real roots – will be *irrational* (they will include *radicals*).

Model Problem 1: *rational roots*

Solve $x^2 + 6x - 7 = 0$ by completing the square.

Solution:

(A) $x^2 + 6x - 7 = 0$

$x^2 + 6x = 7$

(B) $\dfrac{b^2}{4a} = \dfrac{6^2}{4(1)} = \dfrac{36}{4} = 9$

$x^2 + 6x + 9 = 7 + 9$

(C) $(x+3)^2 = 16$

(D) $x + 3 = \sqrt{16}$

$x + 3 = \pm 4$

(E) $x = -3 \pm 4$

$\{1, -7\}$

Explanation of steps:

(A) *[a = 1 is a perfect square.]* Move the constant term *c* to the opposite side *[by adding 7 to both sides]*.

(B) Find $\dfrac{b^2}{4a}$ and add it *[9]* to both sides of the equation.

(C) Factor the trinomial into a binomial squared.
$[x^2 + 6x + 9 = (x+3)^2]$.

(D) Take the square root of both sides. Use the $\pm$ symbol on the right side and simplify radicals. $[\sqrt{16} = \pm 4]$

(E) Solve for *x*, remembering that $\pm$ gives two solutions $[-3+4 = 1 \text{ and } -3-4 = -7]$.

Practice Problems

1. Solve $x^2 + 10x - 11 = 0$ by completing the square.	2. Solve $x^2 - 8x + 16 = 0$ by completing the square.

Model Problem 2: *irrational roots*

Solve $x^2 - 2x - 1 = 0$ by completing the square.

Solution:

(A)
$$x^2 - 2x - 1 = 0$$
$$x^2 - 2x = 1$$

(B)
$$\frac{b^2}{4a} = \frac{(-2)^2}{4(1)} = \frac{4}{4} = 1$$
$$x^2 - 2x + 1 = 1 + 1$$

(C) $(x-1)^2 = 2$

(D) $x - 1 = \pm\sqrt{2}$

(E) $x = 1 \pm \sqrt{2}$
$$\{1 + \sqrt{2}, \ 1 - \sqrt{2}\}$$

Explanation of steps:

(A) *[a = 1 is a perfect square.]*
Move the constant term c to the opposite side.

(B) Add $\dfrac{b^2}{4a}$ to both sides of the equation
[add 1 to both sides].

(C) Factor the left side into a binomial squared.
[$x^2 - 2x + 1 = (x-1)^2$].

(D) Take the square root of both sides. Use the $\pm$ symbol on the right side and simplify radicals.
[$\pm\sqrt{2}$ cannot be simplified.]

(E) Solve for x, remembering that $\pm$ gives two solutions
[we are left with two irrational roots].

Practice Problems

3. Solve $x^2 + 4x + 4 = 0$ by completing the square.	4. Solve $x^2 - 4x - 8 = 0$ by completing the square.
5. Use completing the square to show that $x^2 - 2x + 3 = 0$ has no real solutions.	6. A rectangular pool has an area of 880 square feet. The length is 10 feet longer than the width. Find the dimensions of the pool, to the *nearest tenth of a foot*.

Model Problem 3: *lead coefficient other than 1*

Solve $2x^2 + 6x - 5 = 0$ by completing the square.

Solution:

$$2x^2 + 6x - 5 = 0$$

(A) $4x^2 + 12x - 10 = 0$

(B) $4x^2 + 12x = 10$

(C) $\dfrac{b^2}{4a} = \dfrac{12^2}{4(4)} = \dfrac{144}{16} = 9$

$$4x^2 + 12x + 9 = 10 + 9$$

(D) $(2x + 3)^2 = 19$

(E) $2x + 3 = \pm\sqrt{19}$

(F) $x = \dfrac{-3 \pm \sqrt{19}}{2}$

$$\left\{ \dfrac{-3 + \sqrt{19}}{2}, \dfrac{-3 - \sqrt{19}}{2} \right\}$$

Explanation of steps:

(A) If a is not a perfect square, multiply the equation by a.
    [Since a = 2, multiply the equation by 2.]
    This results in new values for a, b, and c.
    [Now, $a = 4$, $b = 12$, and $c = -10$.]

(B) Move the constant term c to the opposite side.

(C) Add $\dfrac{b^2}{4a}$ to both sides of the equation *[add 9]*.

(D) Factor the trinomial into a binomial squared.

(E) Take the square root of both sides. Use the $\pm$ symbol on the right side and simplify radicals.

    [$\pm\sqrt{19}$ cannot be simplified.]

(F) Solve for x, remembering that $\pm$ gives two solutions.

Practice Problems

7. Solve $4x^2 - 4x - 5 = 0$ by completing the square.	8. Solve $4x^2 + 8x = 45$ by completing the square.
9. Solve $3x^2 - 2x - 1 = 0$ by completing the square.	10. Solve $5x^2 + 10x - 25 = 0$ by completing the square. *(Hint: in this case it's easier to divide the equation by the lead coefficient.)*

REGENTS QUESTIONS

Multiple Choice

1. If $x^2 + 2 = 6x$ is solved by completing the square, an intermediate step would be
   - (1) $(x+3)^2 = 7$
   - (2) $(x-3)^2 = 7$
   - (3) $(x-3)^2 = 11$
   - (4) $(x-6)^2 = 34$

2. Brian correctly used a method of completing the square to solve the equation $x^2 + 7x - 11 = 0$. Brian's first step was to rewrite the equation as $x^2 + 7x = 11$. He then added a number to both sides of the equation. Which number did he add?
   - (1) $\dfrac{7}{2}$
   - (2) $\dfrac{49}{4}$
   - (3) $\dfrac{49}{2}$
   - (4) 49

3. Which equation has the same solution as $x^2 - 6x - 12 = 0$?
   - (1) $(x+3)^2 = 21$
   - (2) $(x-3)^2 = 21$
   - (3) $(x+3)^2 = 3$
   - (4) $(x-3)^2 = 3$

Constructed Response

4. Solve $2x^2 - 12x + 4 = 0$ by completing the square, expressing the result in simplest radical form.

Quadratic Formula and the Discriminant

Key Terms and Concepts

Another method for solving quadratic equations of the form $ax^2 + bx + c = 0$ is by the use of the

quadratic formula, $x = \dfrac{-b \pm \sqrt{b^2 - 4ac}}{2a}$.

(This formula is included in the Reference Sheet at the back of the Regents exam.)

The formula is derived by completing the square on the general equation, $ax^2 + bx + c = 0$.

The $\pm$ **symbol** in the formula allows for the possibility of two solutions (*roots*). Part of the formula is under a square root symbol; this part, $b^2 - 4ac$, is called the *discriminant*.

The **discriminant** $b^2 - 4ac$ tells us about the **number and nature of the roots**.
- If the **discriminant is negative**, then the formula includes the square root of a negative number, so there are **no real roots**. *(You will learn about imaginary roots in Algebra II.)*
- If the **discriminant is zero**, the square root term "disappears" from the formula
  ($\sqrt{0} = 0$), leaving just $x = \dfrac{-b}{2a}$. *(You should recognize this as the equation for the axis of symmetry.)* Therefore, the equation has only **one distinct real root**. As long as a and b are both rational, the root will be rational.
  *(Note: Since a quadratic generally has two roots, some people prefer to say that when the discriminant is zero, there are **two equal roots**.)*
- If the **discriminant is positive**, the $\pm$ symbol before the radical causes **two different real roots** to be produced by the formula.
  - If the discriminant is a *perfect square*, the radical sign will be eliminated, meaning the *two roots are rational* (assuming a and b are rational), which means the equation could have been factored over the integers.
  - If the discriminant is *not a perfect square*, the radical sign cannot be eliminated, so the *two roots are irrational*.

We can relate these situations to the graphs of the corresponding parabolas. The discriminant determines how many *x*-intercepts (*roots*) there are.
For example:

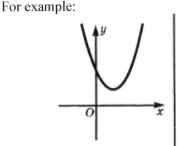 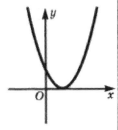 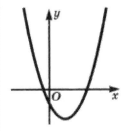

discriminant negative   discriminant zero   discriminant positive
no real roots   one real root   two real roots
(ie, imaginary roots)   *(ie, equal roots)*   *(ie, unequal roots)*

342

Model Problem

In the diagram below, a smaller rectangle with a length of x and width of 2 is enclosed inside a larger rectangle with a length of $2x+3$ and a width of x. The shaded area is 21 square units. Find the length of the smaller rectangle.

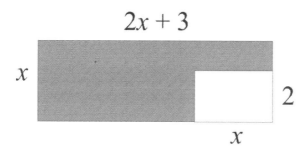

Solution:

(A)   area of large rectangle – area of small rectangle = shaded area

(B)

$$x(2x+3)-2x=21$$
$$2x^2+3x-2x=21$$
$$2x^2+x=21$$
$$2x^2+x-21=0$$

(C)

$$x=\frac{-b\pm\sqrt{b^2-4ac}}{2a}=\frac{-1\pm\sqrt{1^2-4(2)(-21)}}{2(2)}$$

$$=\frac{-1\pm\sqrt{1+168}}{4}=\frac{-1\pm\sqrt{169}}{4}=\frac{-1\pm13}{4}$$

$$x=\frac{12}{4}=3 \quad\text{or}\quad x=\frac{-14}{4}=-3.5 \ (\textit{rejected})$$

The length of the smaller rectangle is 3 units.

Explanation of steps:

(A) Write an equation for the shaded area.

(B) Substitute expressions for the areas (products of the lengths and widths) of the rectangles. Simplify and express in standard form.

(C) Substitute a, b and c [*2, 1, and –21*] into the quadratic formula, and evaluate. Reject any negative roots, since the length cannot be negative.

Practice Problems

1. A quadratic function has two real roots. How many times does the graph of the function cross the *x*-axis? (a) 1 (b) 0 (c) 2 (d) Cannot be determined	2. The roots of $x^2 + 4x + 7$ are (a) not real (imaginary) (b) equal (c) rational (d) irrational
3. Solve $x^2 + 7x + 8 = 0$ by the quadratic formula.	4. Solve $2x^2 - 8x + 3 = 0$ by the quadratic formula.

REGENTS QUESTIONS

Multiple Choice

1. The roots of the equation $2x^2 + 5x - 6 = 0$ are
   - (1) rational and unequal
   - (2) rational and equal
   - (3) irrational and unequal
   - (4) imaginary

2. The roots of the equation $2x^2 + 7x - 3 = 0$ are
   - (1) $-\frac{1}{2}$ and -3
   - (2) $\frac{1}{2}$ and 3
   - (3) $\dfrac{-7 \pm \sqrt{73}}{4}$
   - (4) $\dfrac{7 \pm \sqrt{73}}{4}$

3. The roots of the equation $9x^2 + 3x - 4 = 0$ are
   - (1) imaginary
   - (2) real, rational, and equal
   - (3) real, rational, and unequal
   - (4) real, irrational, and unequal

4. The roots of the equation $x^2 - 10x + 25 = 0$ are
   - (1) imaginary
   - (2) real and irrational
   - (3) real, rational, and equal
   - (4) real, rational, and unequal

5. The discriminant of a quadratic equation is 24. The roots are
   - (1) imaginary
   - (2) real, rational, and equal
   - (3) real, rational, and unequal
   - (4) real, irrational, and unequal

6. The roots of the equation $2x^2 + 4 = 9x$ are
   - (1) real, rational, and equal
   - (2) real, rational, and unequal
   - (3) real, irrational, and unequal
   - (4) imaginary

7. For which value of k will the roots of the equation $2x^2 - 5x + k = 0$ be real and rational numbers?
   - (1) 1
   - (2) −5
   - (3) 0
   - (4) 4

8. What are the roots of the equation $x^2 + 4x - 16 = 0$?
   - (1) $2 \pm 2\sqrt{5}$
   - (2) $-2 \pm 2\sqrt{5}$
   - (3) $2 \pm 4\sqrt{5}$
   - (4) $-2 \pm 4\sqrt{5}$

Constructed Response

9. Solve the equation $6x^2 - 2x - 3 = 0$ and express the answer in simplest radical form.

Word Problems – Quadratic Equations

Key Terms and Concepts

Some verbal problems will require writing and solving quadratic equations. As with any type of verbal problem, try to represent the situation by writing an equation (or system of equations). If the equation is quadratic (in at least one term the variable is squared), then use the methods for solving a quadratic equation.

Geometric area problems often result in quadratic equations.

For example: If the length and width of a rectangle are expressions in terms of the same variable, say x, then the area, as the product of these expressions, would include that variable squared (x^2).

Important: If either root of the equation is not possible in the given situation, reject it.
For example: When finding the length of a side of a rectangle, reject any negative solutions.

Model Problem
Find two consecutive whole numbers such that their product is 42.

Solution:
(A) Let x represent the smaller number. The larger number is $x + 1$.

(B) $x(x+1) = 42$

(C) $x^2 + x = 42$

(D) $x^2 + x - 42 = 0$

(E) $(x+7)(x-6) = 0$

(F) $x + 7 = 0 \ \ or \ \ x - 6 = 0$

(G) $x = -7 \ \ \ or \ \ \ x = 6$

(H) The numbers are 6 and 7.

Explanation of Steps:
(A) Represent an unknown quantity as a variable and express other quantities in terms of this variable. *[Two consecutive integers can be represented by x and x + 1.]*
(B) Write an equation for the given situation.
(C) Simplify both sides of the equation.
(D) Since the equation is quadratic *[there's a x^2 term]*, transform it into standard form by getting zero to one side *[by subtracting 42 from both sides]*.
(E) Factor completely.
(F) Set each factor that contains a variable equal to 0.
(G) Solve each resulting equation. Reject any impossible solutions.
    [The problem asked for whole numbers, so –7 is not a possible solution].
(H) Be sure to answer the problem.
    [If x, the smaller number, is 6, then the larger, next consecutive whole number is 7.]

346

Practice Problems

1. The square of a positive number decreased by twice the number is 48. Find the number.	2. The larger of two positive numbers is 8 more than the smaller. The sum of their squares is 104. Find the two numbers.
3. The area of the rectangular playground enclosure at South School is 500 square meters. The length of the playground is 5 meters longer than the width. Find the dimensions of the playground, in meters.	4. Tamara has two sisters. One of the sisters is 7 years older than Tamara. The other sister is 3 years younger than Tamara. The product of Tamara's two sisters' ages is 24. How old is Tamara?
5. Find two negative consecutive odd integers such that their product is 63.	6. Find three consecutive odd integers such that the product of the first and the second exceeds the third by 8.
7. Find two consecutive whole numbers where the product of the larger and 10 more than the smaller is 90.	8. Three brothers have ages that are consecutive even integers. The product of the first and third boys' ages is 20 more than twice the second boy's age. Find the age of *each* of the three boys.

REGENTS QUESTIONS

Multiple Choice

1. The length of a rectangular window is 5 feet more than its width, w. The area of the window is 36 square feet. Which equation could be used to find the dimensions of the window?

   (1) $w^2 + 5w + 36 = 0$ (3) $w^2 - 5w + 36 = 0$

   (2) $w^2 - 5w - 36 = 0$ (4) $w^2 + 5w - 36 = 0$

2. A rectangle has an area of 24 square units. The width is 5 units less than the length. What is the length, in units, of the rectangle?

   (1) 6   (3) 3

   (2) 8   (4) 19

3. When 36 is subtracted from the square of a number, the result is five times the number. What is the positive solution?

   (1) 9   (3) 3

   (2) 6   (4) 4

4. The length of a rectangle is 3 inches more than its width. The area of the rectangle is 40 square inches. What is the length, in inches, of the rectangle?

   (1) 5   (3) 8.5

   (2) 8   (4) 11.5

5. Byron is 3 years older than Doug. The product of their ages is 40. How old is Doug?

   (1) 10   (3) 5

   (2) 8   (4) 4

6. Noj is 5 years older than Jacob. The product of their ages is 84. How old is Noj?

   (1) 6   (3) 12

   (2) 7   (4) 14

Constructed Response

ength of a rectangle is three times its width. If the width is decreased by 1 inch and ngth is increased by 3 inches, the area of the new rectangle will be 72 square inches. he dimensions, in inches, of the original rectangle.

are and a rectangle have the same area. The length of the rectangle is three inches than the side of the square. The width of the rectangle is two inches less than the f the square. Algebraically, find the side of the square.

9. In the accompanying diagram, the large rectangle *ABCD* is made up of four smaller rectangles with dimensions as indicated.

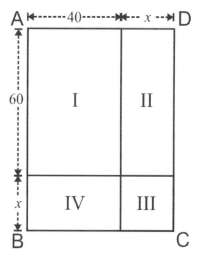

(*a*) Represent, in terms of *x*, the area of *ABCD*.
(*b*) Find the area of *each* of rectangles I, II, III, and IV.
(*c*) Show that the area obtained in part (*a*) above is equal to the sum of the areas obtained in part (*b*) above.

10. Algebraically, find two consecutive positive integers such that the square of the smaller increased by 3 times the larger is equal to 57.

11. The sides of a rectangle are *x* and *x* + 6. The area of the rectangle is 55. Algebraically, find the lengths of the sides.

12. Algebraically, find three consecutive positive integers such that the square of the first is equal to the third.

13. The sum of the squares of two positive consecutive odd integers is 130. Algebraically, find the integers.

14. Barb pulled the plug in her bathtub and it started to drain. The amount of water in the bathtub as it drains is represented by the equation $L = -5t^2 - 8t + 120$, where *L* represents the number of liters of water in the bathtub and *t* represents the amount of time, in minutes, since the plug was pulled.

How many liters of water were in the bathtub when Barb pulled the plug? Show your reasoning.

Determine, to the *nearest tenth of a minute*, the amount of time it takes for all the water in the bathtub to drain.

15. A contractor needs 54 square feet of brick to construct a rectangular walkway. The length of the walkway is 15 feet more than the width. Write an equation that could be used to determine the dimensions of the walkway. Solve this equation to find the length and width, in feet, of the walkway.

16. Find three consecutive positive even integers such that the product of the second and third integers is twenty more than ten times the first integer.

17. As shown in the accompanying diagram, the hypotenuse of the right triangle is 6 meters long. One leg is 1 meter longer than the other. Find the lengths of *both* legs of the triangle, to the *nearest hundredth of a meter*.

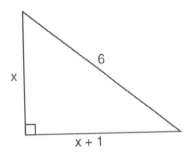

18. A rectangular garden measuring 12 meters by 16 meters is to have a walkway installed around it with a width of *x* meters, as shown in the diagram below. Together, the walkway and the garden have an area of 396 square meters.

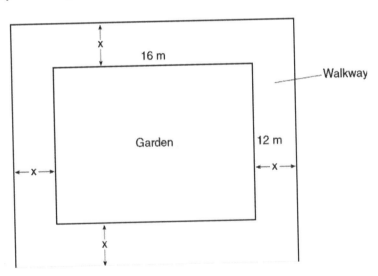

Write an equation that can be used to find *x*, the width of the walkway.

Describe how your equation models the situation.

Determine and state the width of the walkway, in meters.

XV.   PARABOLAS

Finding Roots Given a Parabolic Graph

Key Terms and Concepts

The **standard form** of a quadratic function is $y = ax^2 + bx + c$ where $a \neq 0$.

For example: $y = x^2 - 4x + 3$

The graph of a quadratic function is a U-shaped curve called a **parabola**.

If $a > 0$, then the parabola "**opens up**" like the letter U, but if $a < 0$, then the parabola "**opens down**" like an upside-down U.

When y = 0, we can find the **roots** or **zeros** algebraically by solving for x.

For example: The roots of $y = x^2 - 4x + 3$ are the solutions to the equation $x^2 - 4x + 3 = 0$.
Factoring, we get $(x-1)(x-3) = 0$, so the roots are 1 and 3.

Graphically, y = 0 for all points along the x-axis. Therefore, the **roots** or **zeros** are the x-coordinates of the points **where the parabola crosses the x-axis**, also called the **x-intercepts**.

For example: Since the roots of the quadratic function $y = x^2 - 4x + 3$ are 1 and 3, the parabola will cross the x-axis at points (1,0) and (3,0).

Model Problem

What are the root(s) of the quadratic equation associated with this graph?

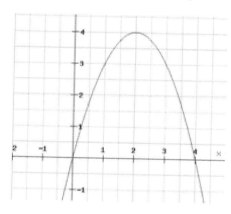

Solution:

The roots are 0 and 4.

Explanation of Steps:

The x-intercepts, or the x-coordinates of the points where the parabola crosses the x-axis, are the roots of the equation. *[Since the parabola crosses the x-axis at (0,0) and (4,0), the roots are 0 and 4. Note that in this grid, each square is one half unit wide.]*

Practice Problems

1. What are the root(s) of the quadratic equation associated with this graph?

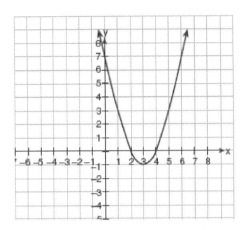

2. What are the root(s) of the quadratic equation associated with this graph?

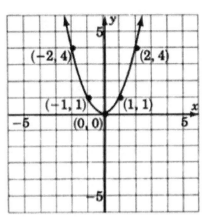

3. What are the root(s) of the quadratic equation associated with this graph?

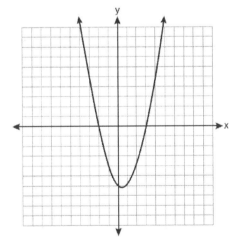

4. What are the root(s) of the quadratic equation associated with this graph?

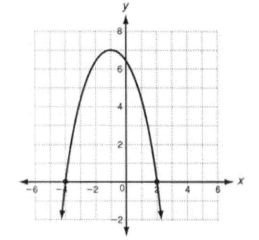

REGENTS QUESTIONS

Multiple Choice

1. The equation $y = x^2 + 3x - 18$ is graphed on the set of axes below.

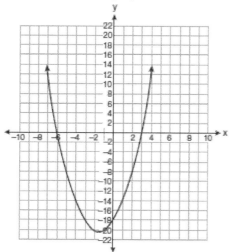

Based on this graph, what are the roots of the equation $x^2 + 3x - 18 = 0$?
    (1) -3 and 6   (3) 3 and -6
    (2) 0 and -18   (4) 3 and 18

2. The equation $y = -x^2 - 2x + 8$ is graphed on the set of axes below.

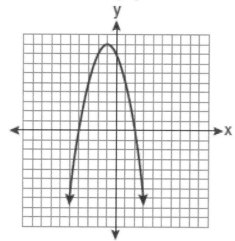

Based on this graph, what are the roots of the equation $-x^2 - 2x + 8 = 0$?
    (1) 8 and 0   (3) 9 and -1
    (2) 2 and -4   (4) 4 and -2

3. A student correctly graphed the parabola shown below to solve a given quadratic equation.

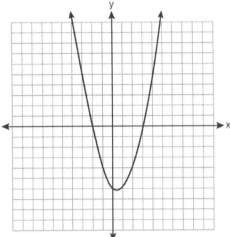

What are the roots of the quadratic equation associated with this graph?
(1) –6 and 3   (3) –3 and 2
(2) –6 and 0   (4) –2 and 3

4. The roots of a quadratic equation can be found using the graph below.

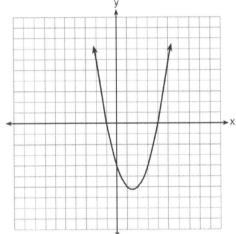

What are the roots of this equation?
(1) –4, only   (3) –1 and 4
(2) –4 and –1   (4) –4, –1, and 4

5. The equation $y = ax^2 + bx + c$ is graphed on the set of axes below.

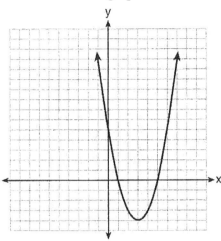

Based on the graph, what are the roots of the equation $ax^2 + bx + c = 0$?
(1) 0 and 5   (3) 1 and 5
(2) 1 and 0   (4) 3 and –4

6. The graphs below represent functions defined by polynomials. For which function are the zeros of the polynomials 2 and –3?

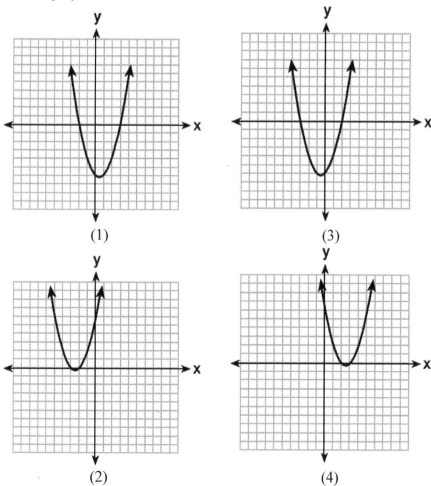

(1)   (3)

(2)   (4)

Finding Vertex and Axis of Symmetry Graphically

Key Terms and Concepts

A **parabola** is a graph of a quadratic function of the form $y = ax^2 + bx + c$ where $a \neq 0$.

The **vertex** of a parabola is the minimum or maximum point, or turning point, on the graph. It is the lowest point on the curve if the parabola opens up ($a > 0$), or the highest point if the parabola opens down ($a < 0$).

The **axis of symmetry** is the vertical line that crosses through the vertex. Its equation is in the form x = *[the x-coordinate of the vertex]*. The axis of symmetry divides the parabola into two parts that are mirror images of each other.

For example: If the vertex of a parabola is (–2, 3), its x-coordinate is –2, so the axis of symmetry is a line whose equation is $x = -2$. In other words, a vertical line drawn from the vertex to the x-axis would cross the axis at –2.

Model Problem
What are the vertex and axis of symmetry of the following parabola?

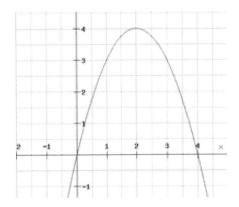

Solution:
  (A) The vertex is (2,4).
  (B) The axis of symmetry is $x = 2$.

Explanation of Steps:
  (A) The vertex is the turning point of the parabola.
  [Since this parabola opens down, the vertex is the highest point, (2,4).]
  (B) The axis of symmetry is a vertical line through the vertex.
  [Given that the vertex is (2,4), a vertical line through this point would include all points where x equals 2, and would cross the x-axis at 2, so the equation of the axis of symmetry is $x = 2$.]

Practice Problems

1. What are the vertex and the axis of symmetry of the parabola shown below?

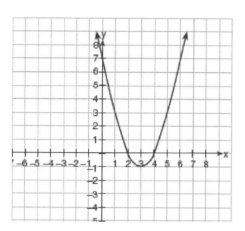

2. What are the vertex and the axis of symmetry of the parabola shown below?

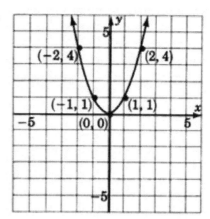

3. What are the vertex and the axis of symmetry of the parabola shown below?

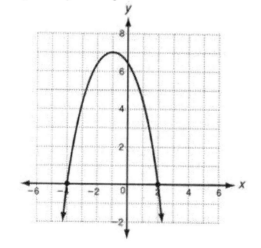

4. What are the vertex and the axis of symmetry of the parabola shown below?

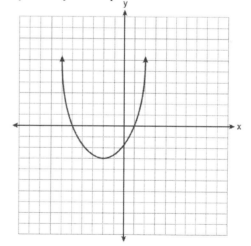

REGENTS QUESTIONS

Multiple Choice

1. What are the vertex and the axis of symmetry of the parabola shown in the diagram below?

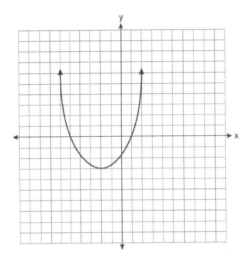

(1) The vertex is (-2,-3), and the axis of symmetry is $x = -2$.
(2) The vertex is (-2,-3), and the axis of symmetry is $y = -2$.
(3) The vertex is (-3,-2), and the axis of symmetry is $y = -2$.
(4) The vertex is (-3,-2), and the axis of symmetry is $x = -2$.

2. A swim team member performs a dive from a 14-foot-high springboard. The parabola below shows the path of her dive.

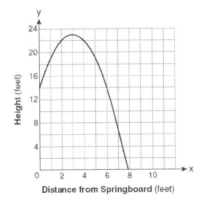

Distance from Springboard (feet)

Which equation represents the axis of symmetry?
(1) $x = 3$ (3) $x = 23$
(2) $y = 3$ (4) $y = 23$

3. If the equation of the axis of symmetry of a parabola is $x = 2$, at which pair of points could the parabola intersect the x-axis?

      (1) (3,0) and (5,0)   (3) (3,0) and (1,0)
      (2) (3,0) and (2,0)   (4) (−3,0) and (−1,0)

4. Which equation represents the axis of symmetry of the graph of the parabola below?

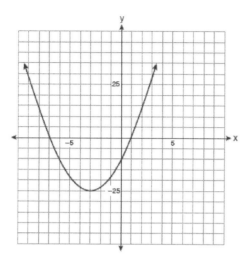

      (1) $y = -3$ (3) $y = -25$
      (2) $x = -3$ (4) $x = -25$

5. What is the equation of the axis of symmetry of the parabola shown in the diagram below?

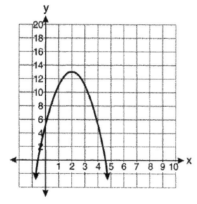

      (1) $x = -0.5$ (3) $x = 4.5$
      (2) $x = 2$ (4) $x = 13$

6. What are the vertex and axis of symmetry of the parabola shown in the diagram below?

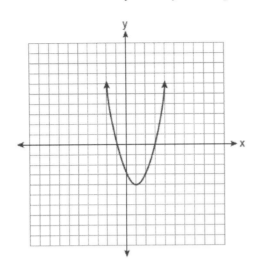

(1) vertex: (1,–4); axis of symmetry: x = 1
(2) vertex: (1,–4); axis of symmetry: x = –4
(3) vertex: (–4,1); axis of symmetry: x = 1
(4) vertex: (–4,1); axis of symmetry: x = –4

7. What are the vertex and the axis of symmetry of the parabola shown in the graph below?

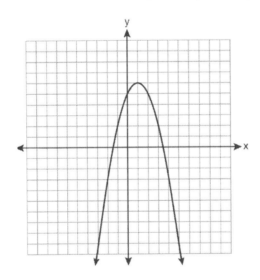

(1) vertex: (1,6); axis of symmetry: $y = 1$
(2) vertex: (1,6); axis of symmetry: $x = 1$
(3) vertex: (6,1); axis of symmetry: $y = 1$
(4) vertex: (6,1); axis of symmetry: $x = 1$

8. What are the coordinates of the vertex and the equation of the axis of symmetry of the parabola shown in the graph below?

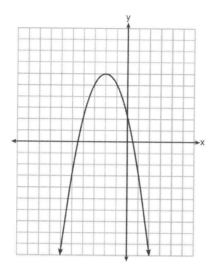

(1) (0,2) and $y = 2$ (3) (–2,6) and $y = -2$

(2) (0,2) and $x = 2$ (4) (–2,6) and $x = -2$

9. Which parabola has an axis of symmetry of $x = 1$?

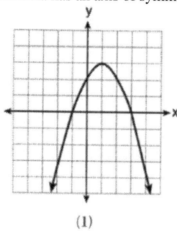

(1)

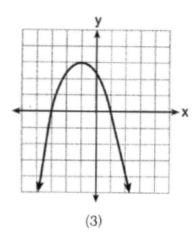

(3)

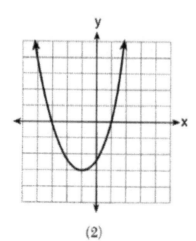

(2)

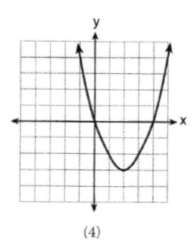

(4)

Constructed Response

10. State the equation of the axis of symmetry and the coordinates of the vertex of the parabola graphed below.

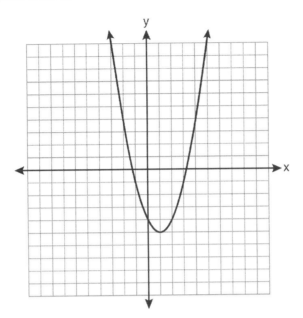

Finding Vertex and Axis of Symmetry Algebraically

Key Terms and Concepts

A **parabola** is a graph of a quadratic function of the form $y = ax^2 + bx + c$ where $a \neq 0$.

For example: For $y = x^2 - 4x + 3$, $a = 1$, $b = -4$, and $c = 3$.

The **axis of symmetry** is a vertical line that divides the parabola into two parts that are mirror images of each other. The **equation for the axis of symmetry** is: $x = \dfrac{-b}{2a}$.

For example: The axis of symmetry for the parabola whose equation is $y = x^2 - 4x + 3$

is the line $x = \dfrac{-b}{2a} = \dfrac{-(-4)}{2(1)} = 2$, or simply $x = 2$.

An alternate method for finding the *equation for the axis of symmetry* is to find the *average of the roots*. This may be easier if the quadratic function is already in factored form. The **factored form** of a quadratic equation is $y = g(x - m)(x - n)$, where m and n are the roots.

For example: If we factor $y = x^2 - 4x + 3$, we get $y = (x - 3)(x - 1)$, so the roots are {3, 1}.

Therefore, the equation for the axis of symmetry is $x = \dfrac{m+n}{2} = \dfrac{3+1}{2} = 2$.

The **vertex** (or turning point) of the parabola is the lowest point (*minimum*) on the curve if the parabola *opens up* ($a > 0$), or the highest point (*maximum*) if the parabola *opens down* ($a < 0$). Since the minimum or maximum point for a parabola is at its vertex, we will often need to find the vertex to determine the largest or smallest value of a real world quadratic function.

The vertex lies on the axis of symmetry. The **x-coordinate of the vertex** is determined by the equation for the axis of symmetry. The **y-coordinate of the vertex** can be found by substituting for x in the quadratic equation.

For example: For $y = x^2 - 4x + 3$, the axis of symmetry is $x = 2$, so substitute 2 for x.

$y = (2)^2 - 4(2) + 3 = -1$, so the vertex is the point (2,–1).

Model Problem 1

Find the axis of symmetry and vertex for the parabola whose equation is $y = x^2 + 12x + 32$.

Solution:

(A) $a = 1$ and $b = 12$

(B) $x = \dfrac{-b}{2a} = \dfrac{-(12)}{2(1)} = -6$

(C) $y = x^2 + 12x + 32$

$y = (-6)^2 + 12(-6) + 32 = -4$

(D) Axis of symmetry: $x = -6$. Vertex: $(-6, -4)$.

Explanation of Steps:

(A) The values for a and b come from the coefficients of the x^2 and x terms of the equation.

(B) Substitute for a and b in the axis of symmetry equation $x = \dfrac{-b}{2a}$ and evaluate.

(C) Substitute the value of x found in the previous step into the original equation to find the value of y. You now have the coordinates of the vertex.

(D) State your answers.

Model Problem 2

You have a 500-ft. roll of chain link fencing and a large field. You want to fence in a rectangular playground area. What is the largest such playground area you can enclose?

Solution:

(A) The perimeter $P = 2l + 2w$, so $500 = 2l + 2w$, or $250 = l + w$.

(B) Therefore, $l = -w + 250$.

(C) The area $A = lw$, so $A = (-w + 250)(w)$, or $A = -w^2 + 250w$.

(D) The axis of symmetry is at $w = \dfrac{-b}{2a} = \dfrac{-250}{2(-1)} = 125$.

(E) When $w = 125$, $l = -w + 250 = -125 + 250 = 125$.

(F) So the maximum area is $A = lw = (125)(125) = 15,625$ sq. ft.

Explanation of Steps:

(A) The amount of fencing tells us the perimeter *[500 ft.]*. Write the formula and substitute.

(B) From the perimeter equation, we can solve for l in terms of w.

(C) Now, we can substitute for l and w into the area formula, resulting in a quadratic function.

(D) Since this quadratic function graphs as a parabola that opens down *[$a = -1$]*, the maximum area is at its vertex. Finding the axis of symmetry will find us the value of w when the area is at its maximum. *[w = 125 ft.]*

(E) Once we have w, we can find l. *[Substitute into the formula from step B.]*

(F) With both dimensions known, calculate the area.

Practice Problems

1. Find the axis of symmetry and vertex of the parabola whose equation is $y = -x^2 + 4x - 8$.	2. Find the axis of symmetry and vertex of the parabola whose equation is $y = x^2 - 6x + 10$.
3. What is the vertex of the parabola whose equation is $y = 3x^2 + 6x - 1$?	4. What is the minimum point of the graph of the equation $y = 2x^2 + 8x + 9$?
5. Find the axis of symmetry and vertex of the parabola whose equation is $y = x^2 + 2x$.	6. Find the axis of symmetry and vertex of the parabola whose equation is $y = 3x^2 + 1$.

REGENTS QUESTIONS

Multiple Choice

1. What are the vertex and axis of symmetry of the parabola $y = x^2 - 16x + 63$?
   (1) vertex: (8,-1); axis of symmetry: $x = 8$
   (2) vertex: (8,1); axis of symmetry: $x = 8$
   (3) vertex: (-8,-1); axis of symmetry: $x = -8$
   (4) vertex: (-8,1); axis of symmetry: $x = -8$

2. The height, y, of a ball tossed into the air can be represented by the equation $y = -x^2 + 10x + 3$, where x is the elapsed time. What is the equation of the axis of symmetry of this parabola?
   (1) $y = 5$ (3) $x = 5$
   (2) $y = -5$ (4) $x = -5$

3. What is an equation of the axis of symmetry of the parabola represented by $y = -x^2 + 6x - 4$?
   (1) $x = 3$ (3) $x = 6$
   (2) $y = 3$ (4) $y = 6$

4. The equation of the axis of symmetry of the graph of $y = 2x^2 - 3x + 7$ is
   (1) $x = \dfrac{3}{4}$ (3) $x = \dfrac{3}{2}$
   (2) $y = \dfrac{3}{4}$ (4) $y = \dfrac{3}{2}$

5. What is the vertex of the parabola represented by the equation $y = -2x^2 + 24x - 100$?
   (1) $x = -6$ (3) (6,–28)
   (2) $x = 6$ (4) (–6,–316)

6. The vertex of the parabola $y = x^2 + 8x + 10$ lies in Quadrant
   (1) I   (3) III
   (2) II   (4) IV

7. What is the vertex of the graph of the equation $y = 3x^2 + 6x + 1$?
   (1) (–1,–2)   (3) (1,–2)
   (2) (–1,10)   (4) (1,10)

Constructed Response

8. Find algebraically the equation of the axis of symmetry and the coordinates of the vertex of the parabola whose equation is $y = -2x^2 - 8x + 3$.

Graphing Parabolas

Key Terms and Concepts

The **standard form** of a quadratic function is $y = ax^2 + bx + c$ where $a \neq 0$.

For example: For $y = x^2 - 4x + 3$, $a = 1$, $b = -4$, and $c = 3$.

The graph of a quadratic function is a U-shaped curve called a **parabola**. If $a > 0$, then the parabola "**opens up**" like the letter U, but if $a < 0$, then the parabola "**opens down**" like an upside-down U.

For example: For $y = x^2 - 4x + 3$, $a = 1$, so the parabola opens up.

We can **graph the parabola** by:
1. drawing the axis of symmetry as a dashed line for reference only
2. plotting the vertex
3. plotting the y-intercept and its reflection
4. plotting any additional points
5. connecting the points with a solid curve

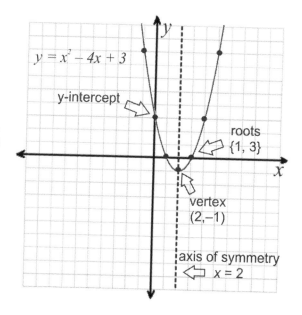

The **equation for the parabola's axis of symmetry**, a vertical line that divides the parabola into two parts that are mirror images of each other, is $x = \dfrac{-b}{2a}$.

For example: The axis of symmetry for the parabola whose equation is

$y = x^2 - 4x + 3$ is the line

$x = \dfrac{-b}{2a} = \dfrac{-(-4)}{2(1)} = 2$, or simply

$x = 2$.

The axis of symmetry gives us the **x-coordinate of the vertex**. You can find **y-coordinate of the vertex** by substituting for x in the quadratic equation.

For example: For $y = x^2 - 4x + 3$, the axis of symmetry is $x = 2$, so substitute 2 for x.
$y = (2)^2 - 4(2) + 3 = -1$, so the vertex is the point (2,–1).

The **y-intercept of the parabola** (the y coordinate of the point where the parabola crosses the y-axis), is c. The **reflection of the y-intercept** over the axis of symmetry (as long as the axis of symmetry is not $x = 0$) is another point on the parabola.

For example: For $y = x^2 - 4x + 3$, $c = 3$, so the y-intercept is 3.
The reflection of (0,3) over the axis of symmetry is (4,3).

We can **plot additional points** on the curve by substituting any integer value of x into the equation to find the corresponding value for y. We can also plot its reflection.

For example: For $y = x^2 - 4x + 3$, we can substitute 5 for x to get $y = (5)^2 - 4(5) + 3 = 8$,
So, the point (5,8) is on the parabola. Its reflection is the point (–1,8).

You can also use the graphing calculator to **graph a parabola**.

For example: To graph $y = x^2 - 4x + 3$,

1. Press $\boxed{Y=}$, then next to "$Y_1 =$", type in the quadratic expression $x^2 - 4x + 3$ using $\boxed{ALPHA}$ [X] for the variable, x, and the $\boxed{x^2}$ button for the exponent, 2.

2. Press $\boxed{ZOOM}$ $\boxed{ZStandard}$ to view the parabola.

The calculator can help you find the **roots** of a graphed parabola:

Using the $\boxed{TABLE}$ feature:

Press $\boxed{2nd}$ $\boxed{TABLE}$ to see a table of (x,y) coordinates of points on the parabola. Press the down arrow key $\boxed{\blacktriangledown}$ to scroll down for more points. You can find the roots by looking for the values of x when **y equals 0** *[for the equation $y = x^2 - 4x + 3$,the roots are 1 and 3].*

Using the $\boxed{CALC}$ feature:

Press $\boxed{2nd}$ $\boxed{CALC}$ $\boxed{zero}$. Look for the first point where the parabola crosses the x-axis. For the "Left Bound?" prompt, use the arrow keys $\boxed{\blacktriangleleft}\boxed{\blacktriangleright}$ to move the cursor to left of this point and press $\boxed{ENTER}$. For the "Right Bound?" prompt, use the arrow keys $\boxed{\blacktriangleleft}\boxed{\blacktriangleright}$ to move the cursor to right of this point (near the vertex) and press $\boxed{ENTER}$ twice. The coordinates are shown; the value of x is the root. If there is a second point where the parabola crosses the x-axis, repeat these steps, moving the cursor to the left and right of the point to find its coordinates; the value of x is another root.

The calculator can also help you find the **vertex** of a graphed parabola:

Press $\boxed{2nd}$ $\boxed{CALC}$ and then select either $\boxed{minimum}$ if the parabola opens up ($a > 0$) or $\boxed{maximum}$ if the parabola opens down ($a < 0$). For the "Left Bound?" prompt, use the arrow keys $\boxed{\blacktriangleleft}\boxed{\blacktriangleright}$ to move the cursor to any point along the left side of the parabola and press $\boxed{ENTER}$. For the "Right Bound?" prompt, use the arrow keys $\boxed{\blacktriangleleft}\boxed{\blacktriangleright}$ to move the cursor to any point along the right side of the parabola and press $\boxed{ENTER}$. For the "Guess?" prompt, press $\boxed{ENTER}$. The coordinates of the vertex (or a close approximation) will be shown. You may need to round to find the actual coordinates of the vertex.

Model Problem

Graph the equation $y = -x^2 + 2x + 3$ on the accompanying set of axes.

Solution:

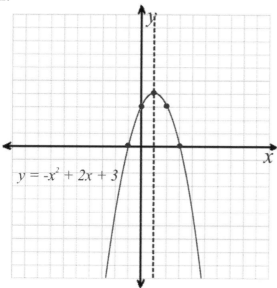

Explanation of Steps:

(A) The axis of symmetry is $x = \dfrac{-b}{2a} = \dfrac{-(2)}{2(-1)} = 1$. A dashed vertical line, $x = 1$, is drawn.

(B) Substituting $x = 1$, $y = -(1)^2 + 2(1) + 3 = 4$, so the vertex is (1, 4).

(C) The y-intercept (c) is 3, so (0,3) and its reflection (2,3) are plotted.

(D) When $x = 3$, $y = -(3)^2 + 2(3) + 3 = 0$, so the point (3,0) and its reflection (−1,0) can be plotted as additional points. Connect the points to draw the parabola.

Practice Problems

1. On separate graph paper, graph the parabola whose equation is $y = 2x^2 - 8x + 4$.	2. On separate graph paper, graph the parabola whose equation is $y = -x^2 + 6x - 5$.
3. On separate graph paper, graph the parabola whose equation is $y = -x^2 + 4x - 8$.	4. On separate graph paper, graph the parabola whose equation is $y = x^2 - 6x + 10$.
5. On separate graph paper, graph the parabola whose equation is $y = 3x^2 + 6x - 1$.	6. On separate graph paper, graph the parabola whose equation is $y = 2x^2 + 8x + 9$.
7. On separate graph paper, graph the parabola whose equation is $y = x^2 + 2x$.	8. On separate graph paper, graph the parabola whose equation is $y = 3x^2 + 1$.

REGENTS QUESTIONS

Constructed Response

1. Graph the equation $y = x^2 - 2x - 3$ on the accompanying set of axes. Using the graph, determine the roots of the equation $x^2 - 2x - 3 = 0$.

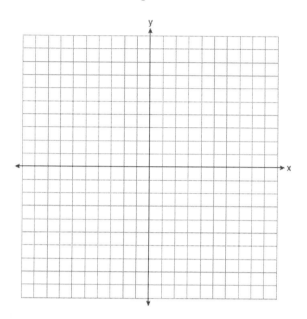

2. On the set of axes below, graph the equation $y = x^2 + 2x - 8$. Using the graph, determine and state the roots of the equation $x^2 + 2x - 8 = 0$.

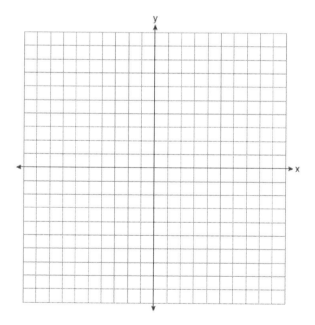

XVI.   QUADRATIC-LINEAR SYSTEMS

Solving Quadratic-Linear Systems Algebraically

Key Terms and Concepts

A system of equations that includes one quadratic equation and one linear equation can be solved algebraically using the **substitution method**.

1. First solve each equation for y so that y is set equal to both a quadratic expression and a linear expression in terms of x.
2. Then, since both expressions are equal to y, set them equal to each other, and solve the resulting quadratic equation for x.
3. Once you have the value(s) of x, find the corresponding value(s) of y by substituting into one of the original equations.

Model Problem

Solve the following system of equations algebraically:

$$y = x^2 - x - 6$$
$$y = 2x - 2$$

Solution:

(A) By substitution, $x^2 - x - 6 = 2x - 2$.

(B) $$x^2 - x - 6 = 2x - 2$$
$$\underline{-2x + 2 \quad -2x + 2}$$
$$x^2 - 3x - 4 = 0$$
$$(x - 4)(x + 1) = 0$$
$$x = \{-1, 4\}$$

(C) When $x = -1$, $y = 2(-1) - 2 = -4$.

When $x = 4$, $y = 2(4) - 2 = 6$.

(D) Solutions: $(-1,-4)$ and $(4,6)$

Explanation of Steps:

(A) Since both expressions are equal to y, set them equal to each other. That is, substitute the linear expression for y in the quadratic equation.

(B) Solve the quadratic equation by getting zero to one side, then factoring.

(C) For each root, substitute for x in the linear equation to find the corresponding value of y.

(D) Express the solutions as ordered pairs.

We can check each solution using the calculator. For example, to check $(-1,-4)$, enter:

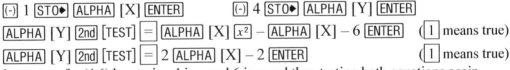

[(-)] 1 [STO▶] [ALPHA] [X] [ENTER]   [(-)] 4 [STO▶] [ALPHA] [Y] [ENTER]

[ALPHA] [Y] [2nd] [TEST] [=] [ALPHA] [X] [x²] − [ALPHA] [X] − 6 [ENTER]   ([1] means true)

[ALPHA] [Y] [2nd] [TEST] [=] 2 [ALPHA] [X] − 2 [ENTER]   ([1] means true)

Repeat these steps for $(4,6)$ by storing 4 in x and 6 in y and then testing both equations again.

Practice Problems

1. Solve the following system of equations algebraically: $$y = x^2 - 5$$ $$y = -4x$$	2. Solve the following system of equations algebraically: $$y = x^2 + 4x + 1$$ $$y = 5x + 3$$
3. Solve the following system of equations algebraically: $$y = x^2 + 2x - 1$$ $$y = 3x + 5$$	4. Solve the following system of equations algebraically: $$y = x^2 + 4x - 2$$ $$y = 2x + 1$$
5. Solve the following system of equations algebraically: $$y = x^2 + 7x + 22$$ $$y + 3x = 1$$	6. Solve the following system of equations algebraically: $$y + 3x = 6$$ $$x^2 = y + 2x + 6$$

REGENTS QUESTIONS

Multiple Choice

1. Which values of x are in the solution set of the following system of equations?
$$y = 3x - 6$$
$$y = x^2 - x - 6$$

   (1) 0, –4   (3) 6, –2
   (2) 0, 4   (4) –6, 2

2. Which ordered pair is a solution to the system of equations $y = x$ and $y = x^2 - 2$?

   (1) (-2,-2)   (3) (0,0)
   (2) (-1,1)   (4) (2,2)

3. Which ordered pair is in the solution set of the system of equations $y = -x + 1$ and
$y = x^2 + 5x + 6$?

   (1) (-5, -1)   (3) (5, -4)
   (2) (-5, 6)   (4) (5, 2)

4. Which ordered pair is a solution of the system of equations $y = x^2 - x - 20$ and
$y = 3x - 15$?

   (1) (-5, -30)   (3) (0, 5)
   (2) (-1, -18)   (4) (5, -1)

5. Which ordered pair is a solution to the system of equations $y = x + 3$ and $y = x^2 - x$?

   (1) (6,9)   (3) (3,–1)
   (2) (3,6)   (4) (2,5)

6. What is the solution set of the system of equations $x + y = 5$ and $y = x^2 - 25$?

   (1) $\{(0,5), (11,-6)\}$ (3) $\{(-5,0), (6,11)\}$
   (2) $\{(5,0), (-6,11)\}$ (4) $\{(-5,10), (6,-1)\}$

7. How many solutions are there for the following system of equations?
$$y = x^2 - 5x + 3$$
$$y = x - 6$$

   (1) 1   (3) 3
   (2) 2   (4) 0

Constructed Response

8. Solve the following system of equations algebraically for all values of x and y.
$$y = x^2 + 2x - 8$$
$$y = 2x + 1$$

Solving Quadratic-Linear Systems Graphically

Key Terms and Concepts

To solve a system of equations graphically, **graph both equations** on the same set of axes and determine the **point(s) of intersection**. These points are the solutions to the system.

For example: The following graph of a quadratic-linear system consists of a **parabola** (from a quadratic equation) and a **line** (from a linear equation). Since the parabola and line intersect at two points, there are two solutions to the system.

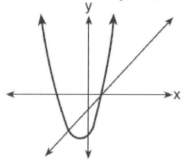

 You can also use the calculator to find the point(s) of intersection:

1. Press $\boxed{Y=}$ and enter both equations.
2. Press $\boxed{2nd}$ $[CALC]$ $\boxed{\text{intersect}}$.
3. Press $\boxed{ENTER}$ for the "First curve?" and "Second curve?" prompts.
4. For the "Guess?" prompt, use the arrow keys to move the cursor near one of the points of intersection. Then press $\boxed{ENTER}$.
5. The coordinates of the closest point of intersection will be shown.
6. If there appears to be a second point of intersection, repeat steps 2 to 5 but move the cursor near the second point in response to the "Guess?" prompt.

Model Problem

Solve the following system of equations graphically:

$$y = x^2 - 4x + 3$$
$$y + 1 = x$$

Solution:

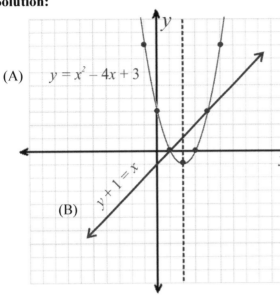

(A) $y = x^2 - 4x + 3$

(B)

(C) Solutions: (1,0) and (4,3)

Explanation of Steps:

(A) Graph the quadratic equation as a parabola. Include at least 3 points with integer values of x on each side of the axis of symmetry.
[The axis of symmetry is $x = 2$ and vertex is (2,–1). Plot additional points with x-coordinates of 3, 4, and 5, plus their reflections.]

(B) Graph the linear equation as a line.
[First transform the equation into slope-intercept form, $y = x - 1$.]

(C) State the point(s) of intersection as the solutions

Practice Problems

1. Solve the following system of equations graphically on separate graph paper: $$y = x^2 + 4x - 2$$ $$y = 2x + 1$$	2. Solve the following system of equations graphically on separate graph paper: $$y = x^2 + 2x - 1$$ $$y = 3x + 5$$
3. Solve the following system of equations graphically on separate graph paper: $$y = x^2 + 4x + 1$$ $$y = 5x + 3$$	4. Solve the following system of equations graphically on separate graph paper: $$y = x^2 + 4x - 1$$ $$y + 3 = x$$

REGENTS QUESTIONS

Multiple Choice

1. Which ordered pair is a solution of the system of equations shown in the graph below?

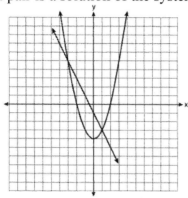

(1) (–3, 1)   (3) (0,–1)
(2) (–3, 5)   (4) (0,–4)

2. Which graph can be used to find the solution of the following system of equations?

$$y = x^2 + 2x + 3$$
$$2y - 2x = 10$$

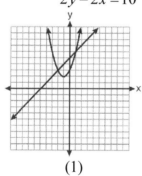

(1)

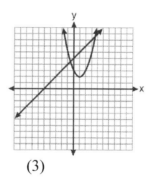

(3)

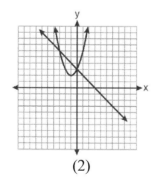

(2)

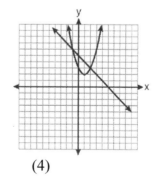

(4)

3. Which graph could be used to find the solution of the system of equations $y = 2x + 6$ and $y = x^2 + 4x + 3$?

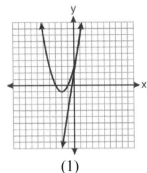

(1)

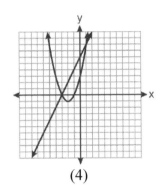

(3)

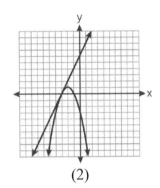

(2)

(4)

4. Two equations were graphed on the set of axes below.

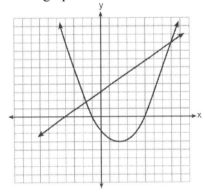

Which point is a solution of the system of equations shown on the graph?

(1) (8, 9)   (3) (0, 3)

(2) (5, 0)   (4) (2, –3)

Constructed Response

5. Solve the following systems of equations graphically, on the set of axes below, and state the coordinates of the point(s) in the solution set.

$$y = x^2 - 6x + 5$$
$$2x + y = 5$$

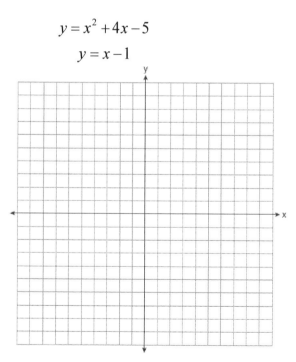

6. On the set of axes below, solve the following system of equations graphically and state the coordinates of all points in the solution set.

$$y = x^2 + 4x - 5$$
$$y = x - 1$$

7. On the set of axes below, solve the following system of equations graphically for all values of x and y.

$$y = x^2 - 6x + 1$$
$$y + 2x = 6$$

8. On the set of axes below, solve the following system of equations graphically for all values of x and y.

$$y = -x^2 - 4x + 12$$
$$y = -2x + 4$$

9. On the set of axes below, solve the following system of equations graphically and state the coordinates of *all* points in the solution set.

$$y = -x^2 + 6x - 3$$
$$x + y = 7$$

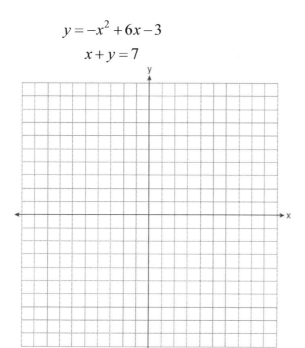

10. On the set of axes below, graph the following system of equations.

$$y + 2x = x^2 + 4$$
$$y - x = 4$$

Using the graph, determine and state the coordinates of *all* points in the solution set for the system of equations.

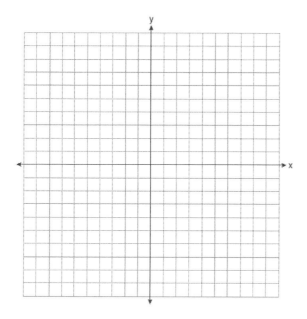

11. On the set of axes below, graph the following system of equations. Using the graph, determine and state *all* solutions of the system of equations.

$$y = -x^2 - 2x + 3$$
$$y + 1 = -2x$$

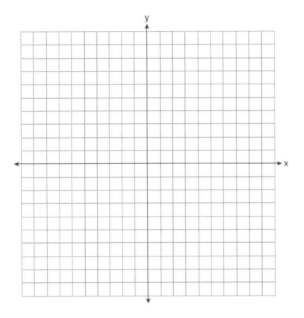

12. On the set of axes below, solve the following system of equations graphically for all values of *x* and *y*. State the coordinates of all solutions.

$$y = x^2 + 4x - 5$$
$$y = 2x + 3$$

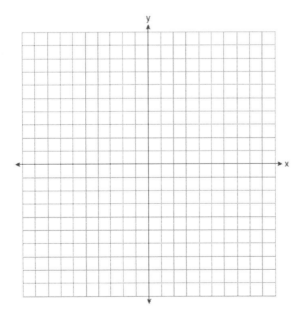

13. A company is considering building a manufacturing plant. They determine the weekly production cost at site A to be $A(x) = 3x^2$ while the production cost at site B is $B(x) = 8x + 3$, where x represents the number of products, *in hundreds*, and $A(x)$ and $B(x)$ are the production costs, *in hundreds of dollars*.

Graph the production cost functions on the set of axes below and label them site A and site B.

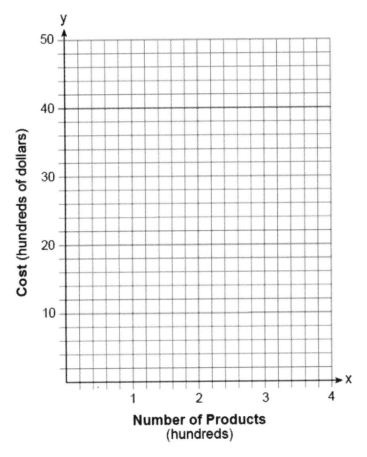

State the positive value(s) of x for which the production costs at the two sites are equal. Explain how you determined your answer.

If the company plans on manufacturing 200 products per week, which site should they use? Justify your answer.

XVII. OTHER FUNCTIONS AND TRANSFORMATIONS

Absolute Value Functions

Key Terms and Concepts

An **absolute value function** can be graphed using a table or a calculator's graphing function.

For example: We can graph $y = |x|$ as follows.

x	$|x|$	y	*(x,y)*
−2	$|-2|$	2	(−2,2)
−1	$|-1|$	1	(−1,1)
0	$|0|$	0	(0,0)
1	$|1|$	1	(1,1)
2	$|2|$	2	(2,2)

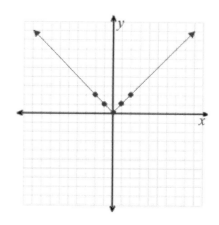

On the calculator, enter: Y= | MATH | NUM | abs(| ALPHA | [X] |) | ZOOM | ZStandard

An absolute value function will generally have a **V shape** (or an upside down V shape).

Model Problem

Use a table to graph the function $y = 2|x+1|$.

Solution:

(A) x	(B) $2|x+1|$	(C) y	(D) *(x,y)*
−2	$2|-2+1|$	2	(−2,2)
−1	$2|-1+1|$	0	(−1,0)
0	$2|0+1|$	2	(0,2)
1	$2|1+1|$	4	(1,4)
2	$2|2+1|$	6	(2,6)

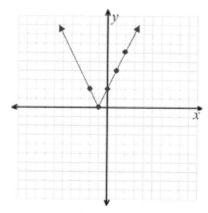

Explanation of Steps:
(A) Pick values of *x* that will evaluate to both positive and negative expressions inside the absolute value sign, allowing you to see both sides of the V shape in the graph.
(B) Substitute the values of *x* into the expression on the right side of the equation.
(C) Evaluate for *y*.
(D) Plot the resulting points on the graph and extend the rays infinitely with arrow heads.

Practice Problems

1. Graph $y = |x| - 3$.

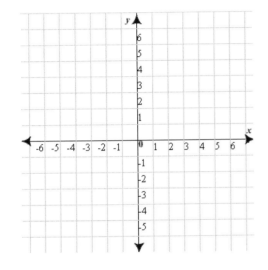

2. Graph $y = -|x|$.

3. Graph $y = 3|x|$.

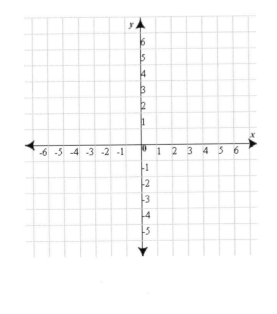

4. Graph $y = \frac{1}{2}|x - 1|$.

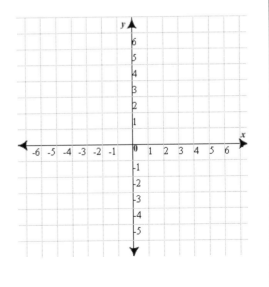

REGENTS QUESTIONS

Multiple Choice

1. Which equation is represented by the graph below?

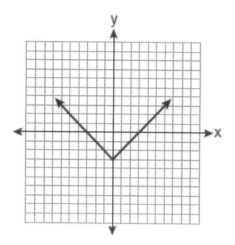

    (1) $y = x^2 - 3$ (3) $y = |x| - 3$

    (2) $y = (x - 3)^2$ (4) $y = |x - 3|$

2. Which is the graph of $y = |x| + 2$?

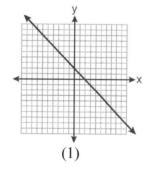

(1)

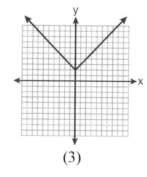

(3)

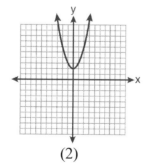

(2)

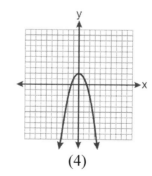

(4)

<u>*Constructed Response*</u>

3. On the set of axes below, graph $y = 2|x + 3|$. Include the interval $-7 \leq x \leq 1$.

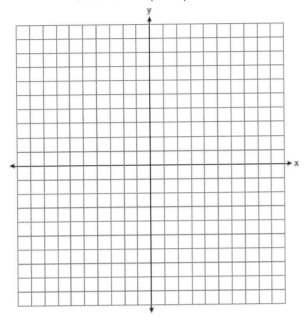

Identifying Families of Functions

Key Terms and Concepts

The graphs of the various types of functions can generally be identified by their shapes.

- A **linear** function is graphed as a **straight line**.
- A **quadratic** function is graphed as a ∪ **or** ∩ **shaped** parabola.
- An **absolute value** function has a ∨ **or** ∧ **shape**.
- An **exponential function** has one of the following shapes:

Model Problem

Identify each of the following graphs by the types of functions:

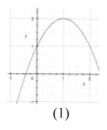   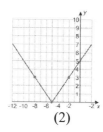   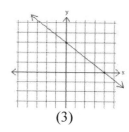   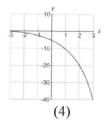

(1)   (2)   (3)   (4)

Solution:

(1) quadratic   (2) absolute value   (3) linear   (4) exponential

Explanation of Steps:

The graphs of functions can be categorized by their shapes.

Practice Problems

1. Which type of function is shown in the accompanying graph? 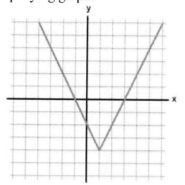	2. Which type of function is shown in the accompanying graph? 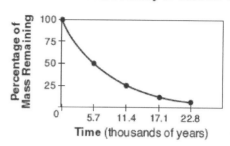
3. Which type of function is shown in the accompanying graph? 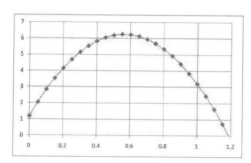	4. Which type of function is shown in the accompanying graph? 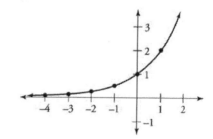
5. Which two types of functions are shown in the accompanying graph? 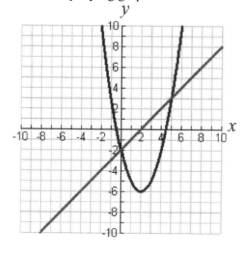	6. Which two types of functions are shown in the accompanying graph? 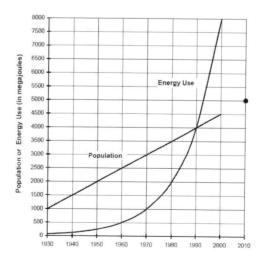

REGENTS QUESTIONS

Multiple Choice

1. Which type of graph is shown in the diagram below?

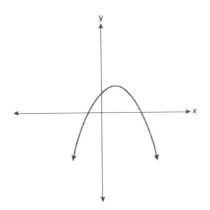

    (1) absolute value   (3) linear
    (2) exponential   (4) quadratic

2. Which graph represents a linear function?

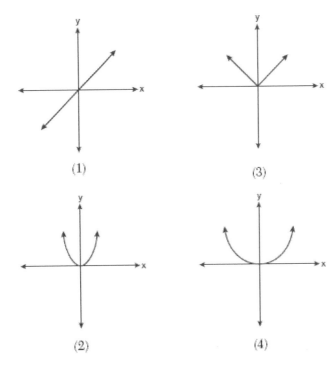

3. Which graph represents an exponential equation?

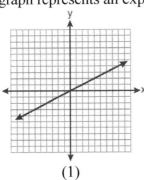

(1)

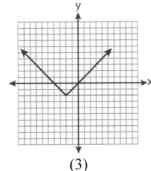

(3)

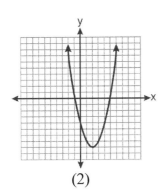

(2)

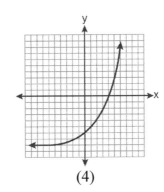

(4)

4. Which type of function is represented by the graph shown below?

(1) absolute value   (3) linear
(2) exponential   (4) quadratic

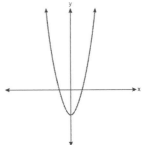

5. Which equation represents a quadratic function?

(1) $y = x + 2$ (3) $y = x^2$

(2) $y = |x + 2|$ (4) $y = 2^x$

6. Which type of function is graphed below?

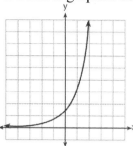

(1) linear   (3) exponential
(2) quadratic   (4) absolute value

7. Which graph represents an absolute value equation?

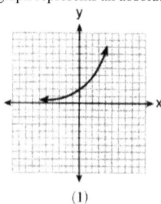

(1)

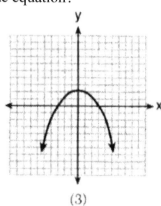

(3)

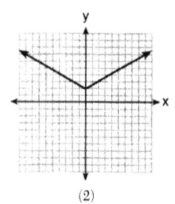

(2)

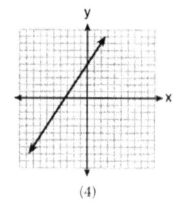

(4)

Cubic, Square Root and Cube Root Functions

Key Terms and Concepts

A **cubic function** is one that is defined by a polynomial with a degree of three; that is, it includes an x^3 term. For the purpose of this course, you need only to be familiar with cubics of the form $y = ax^3 + c$.

The graph of the simplest (*parent*) cubic function, $y = x^3$, is shown below.

x	y
-2	-8
-1	-1
0	0
1	1
2	8

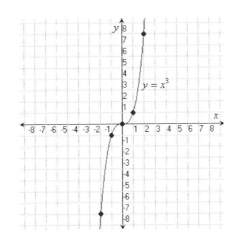

A **square root function** is a function that has the independent variable, *x*, in the radicand.
For examples: $y = \sqrt{x} + 2$ or $y = \sqrt{x-3}$

The simplest (*parent*) square root function, $y = \sqrt{x}$, can be graphed as follows:

x	y
0	0
1	1
4	2
9	3
16	4

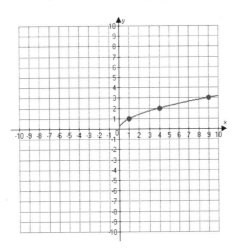

Note that for real numbers, the radicand cannot be negative. This is because there are no real numbers that have negative squares: the square of a positive number is positive and the square of a negative number is positive. So, the domain of a square root function is restricted to only values of *x* for which the radicand is at least zero.

A **cube root function** is similar in definition except that a cube root symbol is used.

For examples: $y = \sqrt[3]{x} + 2$ or $y = \sqrt[3]{x} - 3$

The **cube root** of a number is a factor whose cube equals that number.

For example: $\sqrt[3]{8} = 2$ because $2 \cdot 2 \cdot 2 = 8$

The simplest (*parent*) cube root function, $y = \sqrt[3]{x}$, can be graphed as follows:

x	y
−8	−2
−1	−1
0	0
1	1
8	2
27	3

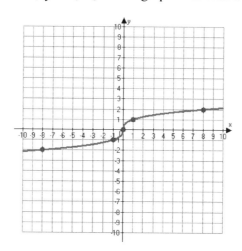

Not that the domain of a cube root function is *not* restricted to non-negative values of x. This is because the cube root of a negative number is a real, negative number.

For example: $\sqrt[3]{-8} = -2$ because $(-2)(-2)(-2) = -8$.

Model Problem

What are the real solutions to the system of equations $y = \sqrt{x}$ and $y = \sqrt[3]{x}$?

Solution: (0,0) and (1,1)

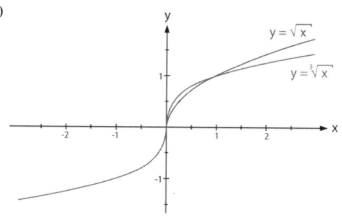

Explanation:

Graph both functions on the same plane and look for intersections.

[These functions intersect at only two points, (0,0) and (1,1). These points of intersection show that $\sqrt{0} = \sqrt[3]{0} = 0$ and $\sqrt{1} = \sqrt[3]{1} = 1$.]

REGENTS QUESTIONS

Constructed Response

1. The number of people, y, involved in recycling in a community is modeled by the function $y = 90\sqrt{3x} + 400$, where x is the number of months the recycling plant has been open. Construct a table of values, sketch the function on the grid, and find the number of people involved in recycling exactly 3 months after the plant opened. After how many months will 940 people be involved in recycling?

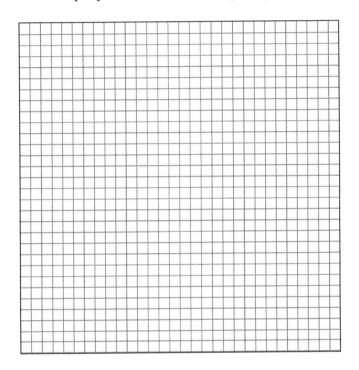

2. Draw the graph of $y = \sqrt{x} - 1$ on the set of axes below.

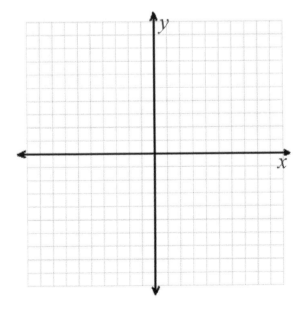

Transformations of Functions

Key Terms and Concepts

The absolute value function $y = |x|$ and the quadratic function $y = x^2$ are graphed below.

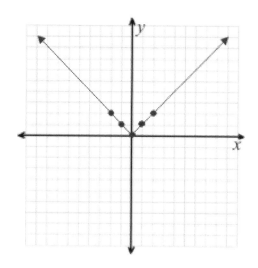 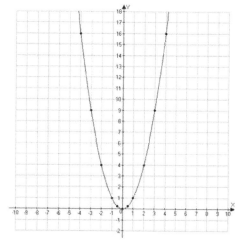

The functions above are called parent functions. A **parent function** is the simplest function of a family of functions that preserves the definition or shape of the entire family.

For example: $y = x^2$ is the parent function for the entire family of quadratic functions that have the general form $y = ax^2 + bx + c$ where $a \neq 0$, since these are all quadratic by definition and are all shaped as parabolas.

The parent functions $y = |x|$ and $y = x^2$ could be transformed in a variety of ways:

$y = |x| + n$ or $y = x^2 + n$ would shift (translate) the function **up** (for a <u>positive n</u>) or **down** (for a <u>negative n</u>) vertically

$y = |x + n|$ or $y = (x + n)^2$ would shift (translate) the function **left** (for a <u>positive n</u>) or **right** (for a <u>negative n</u>) horizontally

$y = -|x|$ or $y = -x^2$ would **flip** (or **reflect**) the graph upside down so that it opens in the opposite direction

$y = n|x|$ or $y = nx^2$ would **vertically stretch** the graph by n and make it more **narrow** (for $n > 1$) or **vertically shrink** the graph and make it more **flat and wide** (for $0 < n < 1$)

$y = |nx|$ or $y = (nx)^2$ would **horizontally shrink** the graph by $\frac{1}{n}$ (for $n > 1$) or **horizontally stretch** the graph by $\frac{1}{n}$ (for $0 < n < 1$)

396

It is often possible to transform other parent functions using similar rules.

For example: The graph of $f(x) = \sqrt{x-a} + b$ can be obtained by translating the graph of the parent function $f(x) = \sqrt{x}$ to a units to the right and then b units up.

Note that this also changes the *domain* and *range*, from $x \geq 0$ and $y \geq 0$ to $x \geq a$ and $y \geq b$, respectively.

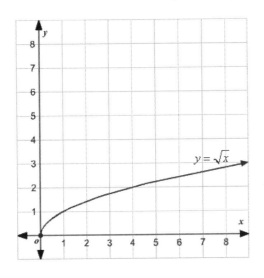
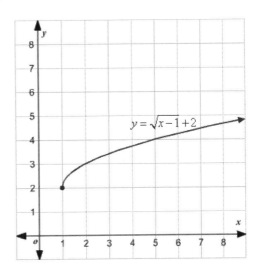

In general, given an original function $f(x)$,

    a) $f(x) + n$ shifts the function graph *up* ($n > 0$) or *down* ($n < 0$)

    b) $f(x + n)$ shifts the function graph *left* ($n > 0$) or *right* ($n < 0$)

    c) $-f(x)$ flips the graph upside down

    d) $n \cdot f(x)$ vertically *stretches* ($n > 1$) or *shrinks* ($0 < n < 1$) the graph by n

    e) $f(nx)$ horizontally *shrinks* ($n > 1$) or *stretches* ($0 < n < 1$) the graph by $\dfrac{1}{n}$

Model Problem

The graph of the function $y = |x|$ is vertically stretched by a factor of 3 and then flipped (reflected) over the *x*-axis. What is the equation of the resulting graph?

Solution:

     (A)   (B)   (C)

     $y = |x|$ $\rightarrow$ $y = 3|x|$ $\rightarrow$ $y = -3|x|$

Explanation of Steps:

    (A) Start with the original equation.

    (B) Perform the first transformation *[stretch by multiplying by the given factor]*.

    (C) Perform the next transformation on the result *[negating will flip it over the x-axis]*.

Practice Problems

1. What is the equation of a graph translated up 3 units, if the original graph is $y = x^2$? (a) $y = (x-3)^2$ (b) $y = (x+3)^2$ (c) $y = x^2 - 3$ (d) $y = x^2 + 3$	2. Describe how the graph $g(x) = (x+2)^2$ is related to the graph of $f(x) = x^2$. (a) a translation 2 units down of $f(x)$ (b) a translation 2 units left of $f(x)$ (c) a translation 2 units up of $f(x)$ (d) a translation 2 units right of $f(x)$
3. If the original graph is $y = |x|$, then the graph of $y = |x-2|$ has been (a) shifted up 2 units (b) shifted down 2 units (c) shifted right 2 units (d) shifted left 2 units	4. If the original graph is $y = x^2$ and the transformed graph is $y = -x^2 - 1$, then the graph has been (a) reflected and shifted down 1 (b) reflected and shifted right 1 (c) dilated and shifted down 1 (d) dilated and shifted right 1
5. To vertically compress the graph of $y = x^2$ by a factor of ½ would result in what new equation? Will the new graph be wider or narrower than the original?	6. If the graph of $y = |x| + 2$ is shifted 3 units down, what is the equation of the new graph?
7. Write the equation of the graph $y = |x|$ after it is shifted 4 units to the left.	8. Write the equation of the graph $y = x^2$ after it is reflected over the x-axis and shifted 1 unit to the right.

REGENTS QUESTIONS

Multiple Choice

1. The diagram below shows the graph of $y = |x - 3|$.

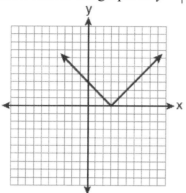

    Which diagram shows the graph of $y = -|x - 3|$?

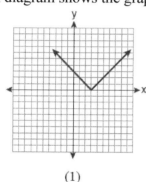

(1)

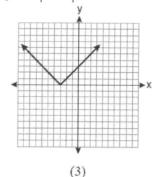

(3)

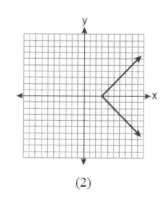

(2)

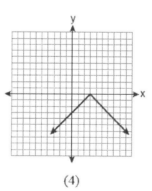

(4)

2. Consider the graph of the equation $y = ax^2 + bx + c$, when $a \neq 0$. If a is multiplied by 3, what is true of the graph of the resulting parabola?
    (1) The vertex is 3 units above the vertex of the original parabola.
    (2) The new parabola is 3 units to the right of the original parabola.
    (3) The new parabola is wider than the original parabola.
    (4) The new parabola is narrower than the original parabola.

3. The graph of the equation $y = |x|$ is shown in the diagram below.

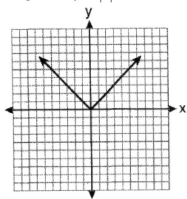

Which diagram could represent a graph of the equation $y = a|x|$ when $-1 < a < 0$?

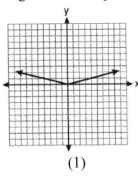

(1)

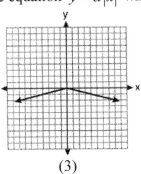

(3)

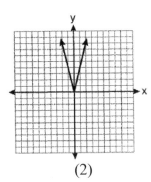

(2)

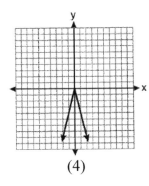

(4)

4. The diagram below shows the graph of $y = -x^2 - c$.

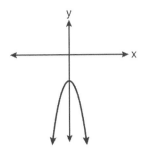

Which diagram shows the graph of $y = x^2 - c$?

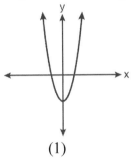

(1)

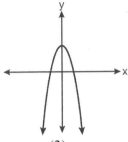

(3)

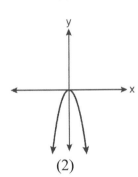

(2)

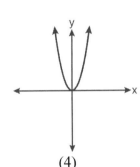

(4)

5. Melissa graphed the equation $y = x^2$ and Dave graphed the equation $y = -3x^2$ on the same coordinate grid. What is the relationship between the graphs that Melissa and Dave drew?

    (1) Dave's graph is wider and opens in the opposite direction from Melissa's graph.
    (2) Dave's graph is narrower and opens in the opposite direction from Melissa's graph.
    (3) Dave's graph is wider and is three units below Melissa's graph.
    (4) Dave's graph is narrower and is three units to the left of Melissa's graph.

6. The graph of $y = |x + 2|$ is shown below.

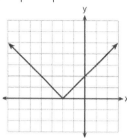

Which graph represents $y = -|x + 2|$?

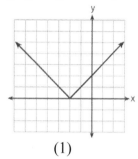

(1)

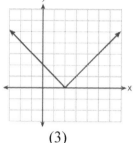

(3)

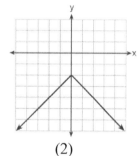

(2)

(4)

7. The graph of a parabola is represented by the equation $y = ax^2$ where a is a positive integer. If a is multiplied by 2, the new parabola will become
    (1) narrower and open downward   (3) wider and open downward
    (2) narrower and open upward   (4) wider and open upward

8. How is the graph of $y = x^2 + 4x + 3$ affected when the coefficient of x^2 is changed to a smaller positive number?
    (1) The graph becomes wider, and the y-intercept changes.
    (2) The graph becomes wider, and the y-intercept stays the same.
    (3) The graph becomes narrower, and the y-intercept changes.
    (4) The graph becomes narrower, and the y-intercept stays the same.

9. Which is the equation of a parabola that has the same vertex as the parabola represented by $y = x^2$, but is wider?
    (1) $y = x^2 + 2$ (3) $y = 2x^2$
    (2) $y = x^2 - 2$ (4) $y = \dfrac{1}{2}x^2$

Constructed Response

10. Graph and label the following equations on the set of axes below.

$$y = |x| \qquad y = \left| \frac{1}{2} x \right|$$

Explain how *decreasing* the coefficient of x affects the graph of the equation $y = |x|$.

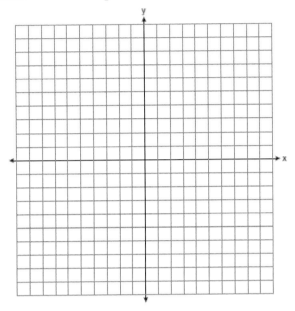

11. On the set of axes below, graph and label the equations $y = |x|$ and $y = 3|x|$ for the interval $-3 \le x \le 3$.

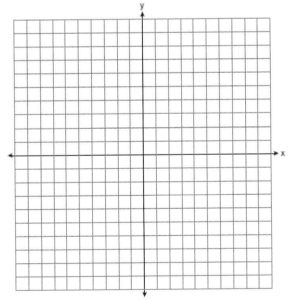

Explain how changing the coefficient of the absolute value from 1 to 3 affects the graph.

12. Graph and label the functions $y = |x|$ and $y = |2x|$ on the set of axes below.

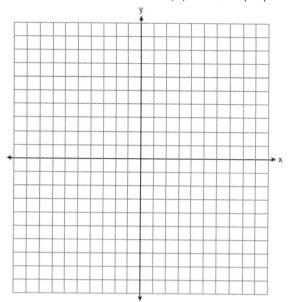

Explain how increasing the coefficient of x affects the graph of $y = |x|$.

13. The vertex of the parabola represented by $f(x) = x^2 - 4x + 3$ has coordinates $(2,-1)$. Find the coordinates of the vertex of the parabola defined by $g(x) = f(x-2)$. Explain how you arrived at your answer.

Piecewise-Defined Functions

Key Terms and Concepts

We often need to define functions in parts. These are called **piecewise-defined functions**. We deifne these functions using two or more pieces, each for a different part of the function's domain, using a large brace symbol, {.

For example: The graph below is made up of two pieces. For values of $x < 0$, we see part of the parabola whose equation is $f(x) = x^2$. But for values of $x \geq 0$ the graph shows the square root function $f(x) = \sqrt{x}$. So, the function is defined as:

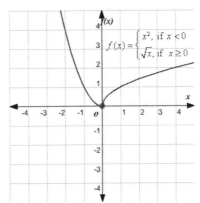

$$f(x) = \begin{cases} x^2 & x < 0 \\ \sqrt{x} & x \geq 0 \end{cases}$$

A **discontinuous** function is a function that is not continuous which, by a loose definition, means that you would not be able to draw or trace the function on a plane without lifting your pencil off the paper. The funcion above is *continuous*, since the two parts meet at and include (0,0), allowing us to draw the graph without lifting our pencil. However, piecewise-defined functions are often discontinuous.

For example: The graph below shows the function defined as pieces of two lines:

$$f(x) = \begin{cases} x+1 & x < 1 \\ x-1 & x \geq 1 \end{cases}$$

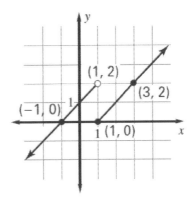

Note that for $f(-1)$ we apply the $x+1$ piece of the definition, $f(-1) = (-1)+1 = 0$, giving the point $(-1,0)$. But for $f(1)$, we apply the $x-1$ piece of the definition, giving $f(1) = 1-1 = 0$ and the point $(1,0)$. Also note that the point $(1,2)$ is an open circle. That is because the top piece of the definition, $f(x) = x+1$, is only for $x < 1$ and not for $x = 1$. When $x \geq 1$, the bottom piece of the definition, $f(x) = x-1$, applies.

405

The **absolute value functions** can also be piecewise defined.

For example: $f(x) = |x|$ can be defined as $f(x) = \begin{cases} -x & x < 0 \\ x & x \geq 0 \end{cases}$

Model Problem
Graph the piecewise-defined function:
$$f(x) = \begin{cases} x^2 & x < 2 \\ 6 & x = 2 \\ 10 - x & 2 < x \leq 6 \end{cases}$$

Solution:

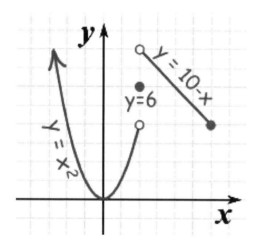

Explanation:
Graph each piece separately, using closed circles at the closed ends of the intervals (where x is =, ≤, or ≥ some value) and open circles at the open ends (where x is < or > some value).

[Graph the top piece as part of the parabola $y = x^2$, but ending at an open circle at (2,4) to show that $x = 2$ ($y = x^2 = 2^2 = 4$) is not part of that piece. The middle piece defines (2,6) as a point on the graph. The bottom piece defines a line segment graphed as $y = 10 - x$ starting at but not including (2,8) – hence the open circle – and ending at and including (6,4).]

406

Practice Problems

1. Given $f(x) = \begin{cases} -x & x < 0 \\ x+1 & x \geq 0 \end{cases}$

   find $f(-3)$, $f(0)$, and $f(2)$.

2. Graph $f(x) = \begin{cases} -x & x < 0 \\ x+1 & x \geq 0 \end{cases}$

   Is this a continuous function?

3. Graph $f(x) = \begin{cases} -1 & x < 1 \\ 1 & x = 1 \\ 2x-2 & x > 1 \end{cases}$

4. Graph $f(x) = \begin{cases} -x+1 & x < 0 \\ 2^x & x \geq 0 \end{cases}$

   Is this a continuous function?

5. Graph $f(x) = |x+1| + 1$. Is this a continuous function? Write an equivalent function using a piecewise definition instead of an absolute value symbol.

6. A garage charges the following rates for parking (with an 8 hour limit):

   $4 per hour for the first 2 hours
   $2 per hour for the next 4 hours
   No charge for the next 2 hours

   Write a piecewise function that gives the parking cost c (in dollars) in terms of the time t (in hours) that a car is parked.

Step Functions

Key Terms and Concepts

A **step function** is a discontinuous function that, when graphed, appears as a series of disconnected line segments resembling steps on a staircase.

Two common step functions are called the floor and ceiling functions. The **floor function** uses special bracket symbols $\lfloor \ \rfloor$ with serifs only at the bottom, which represents the greatest integer less than or equal to the value inside the brackets. The **ceiling function** uses similar bracket symbols $\lceil \ \rceil$ but with serifs only at the top, which represents the least integer that is greater than or equal to the value inside the brackets. For examples: (a) For the floor function $f(x) = \lfloor x \rfloor$, $f(2.9) = 2$ but $f(3) = 3$.

(b) For the ceiling function $g(x) = \lceil x \rceil$, $g(5.3) = 6$ and $g(6) = 6$.

The graphs of floor and ceiling functions are shown below. The graph is made up of disconnected line segments with open or closed circles at their ends. Just as we saw when graphing inequalities, an **open circle** means the point is *excluded*, but a **closed circle** means the point is *included*.

For example: On the graph of $y = \lfloor x \rfloor$ to the left below, $y = \lfloor x \rfloor = 1$ for all real values of x between 1 and 2, including 1 but excluding 2. In other words, $y = 1$ for all values of x such that $1 \le x < 2$, even for values very close to 2 such as 1.99999, but not including 2. For $x = 2$, $y = \lfloor 2 \rfloor = 2$, so the point $(2,2)$ is closed while the point $(2,1)$ is open.

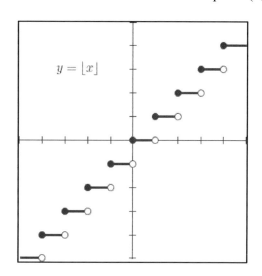
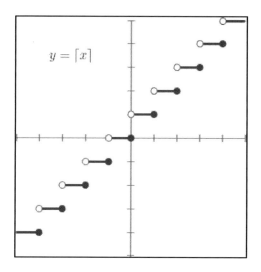

Step functions are used in various real world situations. For example, the graph below shows the cost of mailing a letter way back in 2006. For letters of up to 1 ounce in weight, the postage cost 39 cents. The cost was 41 cents for letters that were more than 1 ounce but up to 2 ounces in weight. For weights of x ounces in the interval $2 < x \leq 3$, the cost was 43 cents. This pattern continued so that each additional ounce, or fraction of an ounce, cost an additional 2 cents.

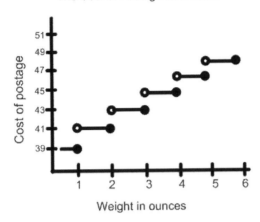

The Cost of Postage for a Letter

This function is a type of ceiling function, defined as $y = 2\lceil x \rceil + 37$ for all $x > 0$. Although it is not continuous, it is still a function, as we can see by applying the vertical line test, below.

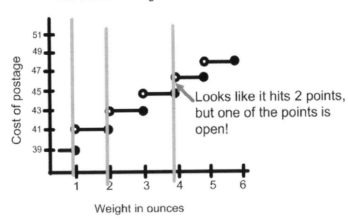

The Cost of Postage for a Letter

Looks like it hits 2 points, but one of the points is open!

Practice Problems

1. Given $f(x) = 3\lceil x \rceil + 5$, find $f(6.25)$.	2. Graph $f(x) = \lfloor x \rfloor + 1$.

XVIII. SAMPLE REGENTS EXAM

The next few pages may be used as a sample exam covering the entire course. Questions were taken from various sections throughout this book, including past Regents Questions. It is designed to look like an actual exam, including all directions, scrap graph paper, the reference sheet, and page numbers that correspond to the "exam booklet" pages.

Unofficial Sample

REGENTS HIGH SCHOOL EXAMINATION

ALGEBRA I (Common Core)

CourseWorkbooks.com

Student Name:_____

School Name: _____

The possession or use of any communications device is strictly prohibited when taking this examination. If you have or use any communications device, no matter how briefly, your examination will be invalidated and no score will be calculated for you.

Print your name and the name of your school on the lines above.

A separate answer sheet for Part I has been provided to you. Follow the instructions from the proctor for completing the student information on your answer sheet.

This examination has four parts, with a total of 37 questions. You must answer all questions in this examination. Record your answers to the Part I multiple-choice questions on the separate answer sheet. Write your answers to the questions in Parts II, III, and IV directly in this booklet. All work should be written in pen, except graphs and drawings, which should be done in pencil. Clearly indicate the necessary steps, including appropriate formula substitutions, diagrams, graphs, charts, etc. The formulas that you may need to answer some questions in this examination are found at the end of the examination. This sheet is perforated so you may remove it from this booklet.

Scrap paper is not permitted for any part of this examination, but you may use the blank spaces in this booklet as scrap paper. A perforated sheet of scrap graph paper is provided at the end of this booklet for any question for which graphing may be helpful but is not required. You may remove this sheet from this booklet. Any work done on this sheet of scrap graph paper will *not* be scored.

When you have completed the examination, you must sign the statement printed at the end of the answer sheet, indicating that you had no unlawful knowledge of the questions or answers prior to the examination and that you have neither given nor received assistance in answering any of the questions during the examination. Your answer sheet cannot be accepted if you fail to sign this declaration.

Notice…
A graphing calculator and a straightedge (ruler) must be available for you to use while taking this examination.

DO NOT OPEN THIS EXAMINATION BOOKLET UNTIL THE SIGNAL IS GIVEN.

Part I

Answer all 24 questions in this part. Each correct answer will receive 2 credits. No partial credit will be allowed. For each statement or question, choose the word or expression that, of those given, best completes the statement or answers the question. Record your answers on your separate answer sheet. [48]

Use this space for computations.

1. The discrete function g is defined as $g(n) = n+1$. If the domain is the set of whole numbers, what is the range?

   (1) $\{0, 1, 2, 3, \dots\}$

   (2) $\{\text{real numbers}\}$

   (3) $\{\dots -2, -1, 0, 1, 2, 3, \dots\}$

   (4) $\{1, 2, 3, 4, \dots\}$

2. The figures below represent the first four terms of a sequence. Assuming the pattern continues, which formula can be used to determine b_n, the number of black triangles in the nth term?

   Figure 1   Figure 2   Figure 3   Figure 4

   (1) $b_n = n+3$

   (2) $b_n = n+6$

   (3) $b_n = 3n$

   (4) $b_n = 3^n$

3. What is the sum of $x^2 - 3x + 7$ and $3x^2 + 5x - 9$?

   (1) $4x^2 - 8x + 2$

   (2) $4x^2 + 2x + 16$

   (3) $4x^2 - 2x - 2$

   (4) $4x^2 + 2x - 2$

4. When adding 2 to the sum, $3+x$, which property allows us to perform the first step below, from line 1 to line 2?

$$[\text{line 1}] \quad 2+(3+x) =$$
$$[\text{line 2}] \quad (2+3)+x =$$
$$[\text{line 3}] \quad 5+x$$

(1) commutative property of addition

(2) addition property of equality

(3) associative property of addition

(4) distributive property of multiplication over addition

Use this space for computations.

5. For which function are –3 and 5 the zeros of the function?

(1) $f(x)=x^2+2x-15$ (3) $f(x)=x^2+2x+15$

(2) $f(x)=x^2-2x-15$ (4) $f(x)=x^2-2x+15$

6. Which value of x is the solution of $\dfrac{x}{3}+\dfrac{x+1}{2}=x$?

(1) 1   (3) 3

(2) –1   (4) –3

7. What is a formula for the nth term of sequence B shown below?
$$B = 10, 12, 14, 16, \ldots$$

(1) $b_n=8+2n$ (3) $b_n=10(2)^n$

(2) $b_n=10n+2$ (4) $b_n=10(2)^{n-1}$

Algebra I (Common Core) – Sample

8. What could be the approximate value of the correlation coefficient for the accompanying scatter plot?

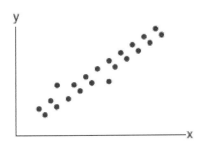

(1) –0.85   (3) 0.21

(2) –0.16   (4) 0.90

9. When $2x^2 - 18$ is factored completely, the result is

(1) $2(x-3)(x+3)$ (3) $(2x+6)(x-3)$

(2) $(2x-6)(x+3)$ (4) $2(x^2-9)$

10. Abbey starts with $20 and plays an arcade game that costs 50 cents per game. Which function could be used to represent the amount of money Abbey has remaining after g games are played?

(1) $f(g)=20g+0.50$ (3) $f(g)=20g-0.50$

(2) $f(g)=0.50g+20$ (4) $f(g)=-0.50g+20$

11. Point $(k,-3)$ lies on the line whose equation is $x-2y=-2$. What is the value of k?

(1) –8   (3) 6

(2) –6   (4) 8

12. The population growth of Boomtown is shown in the accompanying graph.

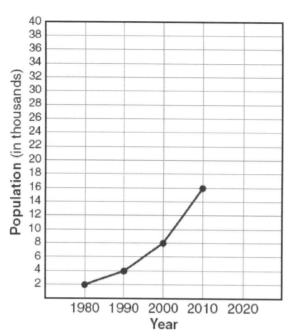

If the same pattern of population growth continues, what will the population of Boomtown be in the year 2020?

(1) 20,000   (3) 40,000

(2) 32,000   (4) 64,000

13. If $x^2 + 2 = 6x$ is solved by completing the square, an intermediate step would be

(1) $(x+3)^2 = 7$ (3) $(x-3)^2 = 11$

(2) $(x-3)^2 = 7$ (4) $(x-6)^2 = 34$

14. Kathy plans to purchase a car that depreciates (loses value) at a rate of 14% per year. The initial cost of the car is \$21,000. Which equation represents the value, v, of the car after 3 years?

(1) $v = 21,000(0.14)^3$ (3) $v = 21,000(1.14)^3$

(2) $v = 21,000(0.86)^3$ (4) $v = 21,000(0.86)(3)$

15. Given the graph of the function $f(x)$ below, for which interval does the function appear to decrease at a constant rate?

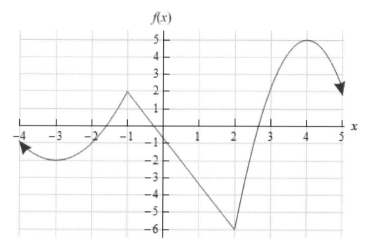

(1) $x < -1$

(3) $2 < x < 4$

(2) $-1 < x < 2$

(4) $x > 4$

16. Which graph does *not* represent a function?

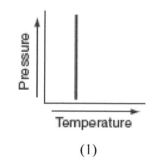

(1)

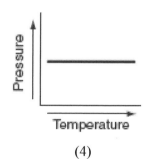

(3)

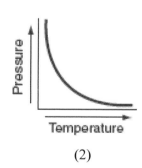

(2)

(4)

17. What is an equation of the line that passes through the points (3,–7) and (3,2)?

(1) $y = \dfrac{5}{3}x$

(3) $x = 3$

(2) $y = -7$

(4) $y = \dfrac{5}{3}x - 12$

18. Jean's scores on five math tests were 98, 97, 99, 98, and 96. Her scores on five English tests were 78, 84, 95, 72, and 79. Which statement is true about the standard deviations for the scores?

(1) The standard deviation for the English scores is greater than the standard deviation for the math scores.

(2) The standard deviation for the math scores is greater than the standard deviation for the English scores.

(3) The standard deviations for both sets of scores are equal.

(4) More information is needed to determine the relationship between the standard deviations.

19. Which inequality is shown on the accompanying graph?

(1) $-4 < x < 2$

(3) $-4 < x \leq 2$

(2) $-4 \leq x < 2$

(4) $-4 \leq x \leq 2$

20. What is the value of y in the following system of equations?

$$2x + 3y = 6$$
$$2x + y = -2$$

(1) 1

(3) –3

(2) 2

(4) 4

21. Which statement correctly describes how the graph of $g(x) = (x+2)^2$ is related to the graph of $f(x) = x^2$?

(1) $g(x)$ is a translation 2 units down from $f(x)$

(2) $g(x)$ is a translation 2 units to the left of $f(x)$

(3) $g(x)$ is a translation 2 units up from $f(x)$

(4) $g(x)$ is a translation 2 units to the right of $f(x)$

Use this space for computations.

22. For which value of k will the roots of the equation $2x^2 - 5x + k = 0$ be rational numbers?

(1) 1

(2) –5

(3) 0

(4) 4

23. The expression $x(x-y)(x+y)$ is equivalent to

(1) $x^2 - y^2$

(2) $x^3 - y^3$

(3) $x^3 - xy^2$

(4) $x^3 - x^2y + y^2$

24. If the equation of the axis of symmetry of a parabola is $x = 2$, at which pair of points could the parabola intersect the x-axis?

(1) (3,0) and (5,0)

(2) (3,0) and (2,0)

(3) (3,0) and (1,0)

(4) (–3,0) and (–1,0)

Part II

Answer all 8 questions in this part. Each correct answer will receive 2 credits. Clearly indicate the necessary steps, including appropriate formula substitutions, diagrams, graphs, charts, etc. For all questions in this part, a correct numerical answer with no work shown will receive only 1 credit. All answers should be written in pen, except for graphs and drawings, which should be done in pencil. [16]

25. Solve for c in terms of a and b: $bc + ac = ab$

26. What is the square root of $4x^2 + 12x + 9$, written as a binomial?

27. Marsha is buying plants and soil for her garden. The soil cost $4 per bag, and the plants cost $10 each. She wants to buy *at least* 5 plants and can spend *no more than* $100.

Write a system of linear inequalities to model the situation. State what each variable in your model represents.

28. Based on data from last Sunday's professional football games, a linear regression equation $y = 0.75x - 0.25$ is created to predict the number of points scored by a team in one game based on the team's time of possession (in minutes) in that game.

How many points, to the *nearest whole number*, would you predict a team to score if their time of possession was 35 minutes?

A team has 35 minutes of possession in a game and scores 32 points. What is the value of the residual for this data point, (35, 32)?

29. A ball is shot straight up in the air from ground level and its height is recorded every 0.5 seconds until it lands 4 seconds later. A graph of the height of the ball over time is shown below. Find the average rate of change, in centimeters per second, over the first 2 second interval.

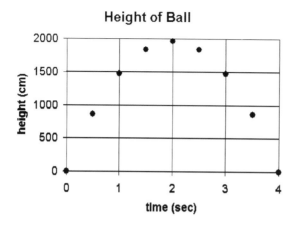

30. The scatter plots below display the same data about the ages of eight health club members and their heart rates during exercises. Which line is a better fit for the data? Explain your reasoning.

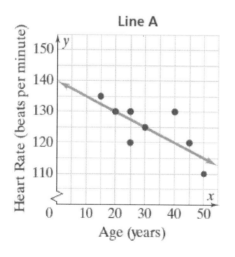

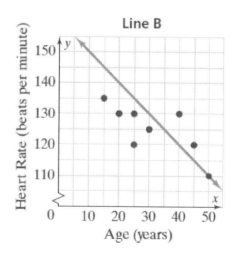

31. The accompanying graph shows Marie's distance from home (A) to work (F) at various times during her drive.

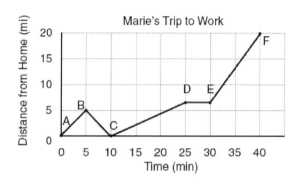

Marie left her briefcase at home and had to return to get it. Which point represents when she turned back around to go home?

Marie also had to wait at the railroad tracks for a train to pass. How long did she wait?

32. The following data values represent test scores on an algebra exam. Find the interquartile range.

$$86, 72, 85, 89, 86, 92, 73, 71, 91, 82$$

Part III

Answer all 4 questions in this part. Each correct answer will receive 4 credits. Clearly indicate the necessary steps, including appropriate formula substitutions, diagrams, graphs, charts, etc. For all questions in this part, a correct numerical answer with no work shown will receive only 1 credit. All answers should be written in pen, except for graphs and drawings, which should be done in pencil. [16]

33. The first table below shows the number of books sold at a library sale.

    Complete this joint frequency table by writing the marginal frequencies in the blank cells.

	Fiction	Nonfiction	Total
Hardcover	28	52	
Paperback	84	36	
Total			

From the same data, complete the relative frequency table below, using *percents*.

	Fiction	Nonfiction	Total
Hardcover			
Paperback			
Total			100%

34. On the grid below, graph the function f over the interval $-10 \le x \le 10$, where f is defined as

$$f(x) = \begin{cases} x+1, & x < 1 \\ x-1, & x \ge 1 \end{cases}$$

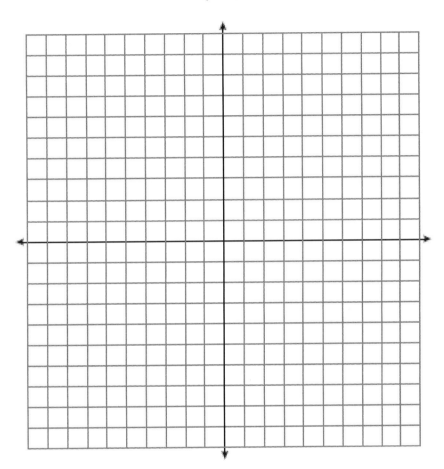

35. Barb pulled the plug in her bathtub and it started to drain. The amount of water in the bathtub as it drains is represented by the equation $L = -5t^2 - 8t + 120$, where L represents the number of liters of water in the bathtub and t represents the amount of time, in minutes, since the plug was pulled.

How many liters of water were in the bathtub when Barb pulled the plug? Show your reasoning.

Determine, to the *nearest tenth of a minute*, the amount of time it takes for all the water in the bathtub to drain.

36. In the accompanying diagram, the large rectangle ABCD is made up of four smaller rectangles with dimensions as indicated.

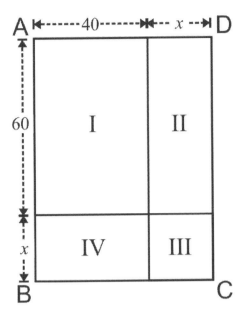

Show algebraically that the area of rectangle ABCD is equal to the sum of the areas of the rectangles I, II, III, and IV.

Part IV

Answer the question in this part. A correct answer will receive 6 credits. Clearly indicate the necessary steps, including appropriate formula substitutions, diagrams, graphs, charts, etc. A correct numerical answer with no work shown will receive only 1 credit. The answer should be written in pen. [6]

37. To raise funds, a club is publishing and selling a calendar. The club has sold $500 in advertising and will sell copies of the calendar for $20 each. The cost of printing each calendar is $6.

   a) In function notation, write $R(x)$ to represent the revenue generated by selling both the advertising and x copies of the calendar. In function notation, write $E(x)$ to represent the expenses for printing x calendars.

   b) Describe how the function $P(x)$, which gives the club's profit on x calendars, is related to $R(x)$ and $E(x)$, and write a function rule for $P(x)$ in simplest form.

   c) What is the minimum number of calendars that would have to be sold in order to generate a profit of at least $2000?

Scrap Graph Paper — This sheet will *not* be scored.

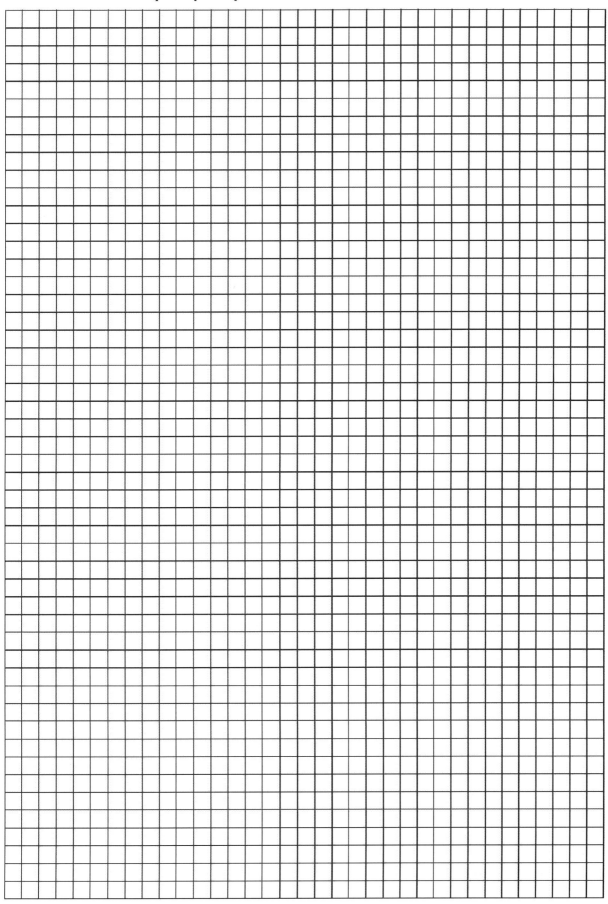

Algebra I (Common Core) – Sample

Scrap Graph Paper — This sheet will *not* be scored.

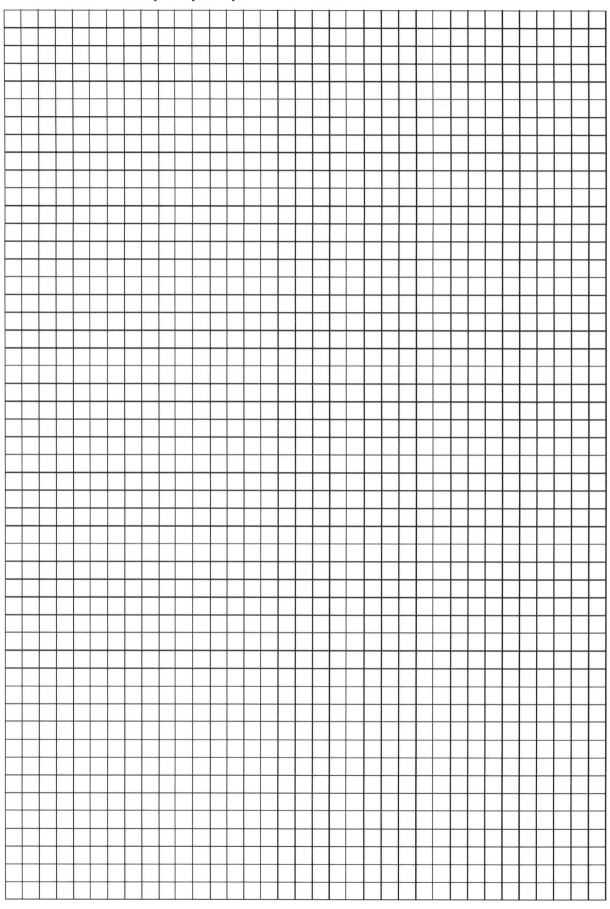

Algebra I (Common Core) – Sample

High School Math Reference Sheet

1 inch = 2.54 centimeters
1 meter = 39.37 inches
1 mile = 5280 feet
1 mile = 1760 yards
1 mile = 1.609 kilometers

1 kilometer = 0.62 mile
1 pound = 16 ounces
1 pound = 0.454 kilogram
1 kilogram = 2.2 pounds
1 ton = 2000 pounds

1 cup = 8 fluid ounces
1 pint = 2 cups
1 quart = 2 pints
1 gallon = 4 quarts
1 gallon = 3.785 liters
1 liter = 0.264 gallon
1 liter = 1000 cubic centimeters

Triangle	$A = \frac{1}{2}bh$
Parallelogram	$A = bh$
Circle	$A = \pi r^2$
Circle	$C = \pi d$ or $C = 2\pi r$
General Prisms	$V = Bh$
Cylinder	$V = \pi r^2 h$
Sphere	$V = \frac{4}{3}\pi r^3$
Cone	$V = \frac{1}{3}\pi r^2 h$
Pyramid	$V = \frac{1}{3}Bh$

Pythagorean Theorem	$a^2 + b^2 = c^2$
Quadratic Formula	$x = \dfrac{-b \pm \sqrt{b^2 - 4ac}}{2a}$
Arithmetic Sequence	$a_n = a_1 + (n-1)d$
Geometric Sequence	$a_n = a_1 r^{n-1}$
Geometric Series	$S_n = \dfrac{a_1 - a_1 r^n}{1 - r}$ where $r \neq 1$
Radians	1 radian = $\dfrac{180}{\pi}$ degrees
Degrees	1 degree = $\dfrac{\pi}{180}$ radians
Exponential Growth/Decay	$A = A_0 e^{k(t-t_0)} + B_0$

Algebra I (Common Core) – Sample

XIX. CORRELATION TO COMMON CORE STANDARDS

N.RN.1-2	*Covered in Algebra 2*

N.RN.3	Explain why the sum or product of two rational numbers is rational; that the sum of a rational number and an irrational number is irrational; and that the product of a nonzero rational number and an irrational number is irrational
	  II. Properties
	  VII. Irrational Numbers

N.Q.1	Use units as a way to understand problems and to guide the solution of multi-step problems; choose and interpret units consistently in formulas; choose and interpret the scale and the origin in graphs and data displays XI. Word Problems – Function Graphs

N.Q.2	Define appropriate quantities for the purpose of descriptive modeling
	  III. Word Problems – Linear Equations
	  III. Word Problems – Inequalities
	  XI. Word Problems – Function Graphs
	  XIV. Word Problems – Quadratic Equations

N.Q.3	Choose a level of accuracy appropriate to limitations on measurement when reporting quantities.
	  I. Rounding
	  I. Precision of Measurements

N.CN	*Covered in Algebra 2 and Precalculus*
N.VM	*Covered in Precalculus*

A.SSE.1	a. Interpret parts of an expression, such as terms, factors, and coefficients
	  VI. POLYNOMIALS
	b. Interpret complicated expressions by viewing one or more of their parts as a single entity
	  XIII. FACTORING

A.SSE.2	Use the structure of an expression to identify ways to rewrite it
	  VI. POLYNOMIALS
	  XIII. FACTORING

A.SSE.3	a. Factor a quadratic expression to reveal the zeros of the function it defines
	  XIII. FACTORING
	b. Complete the square in a quadratic expression to reveal the maximum or minimum value of the function it defines
	  XIV. Completing the Square

A.SSE.4	*Covered in Algebra 2*
A.APR.1	Understand that polynomials form a system analogous to the integers, namely, they are closed under the operations of addition, subtraction, and multiplication; add, subtract, and multiply polynomials VI. POLYNOMIALS
A.APR.2	*Covered in Algebra 2*

A.APR.3	Identify zeros of polynomials when suitable factorizations are available, and use the zeros to construct a rough graph of the function defined by the polynomial.
	    XIII. FACTORING
	    XIV. Finding Quadratic Equations from Given Roots
	    XV. PARABOLAS

A.APR.4-7	*Covered in Algebra 2 and Precalculus*

A.CED.1	Create equations and inequalities in one variable and use them to solve problems
	    III. Word Problems – Linear Equations
	    III. Word Problems – Inequalities
	    XIV. Word Problems – Quadratic Equations

A.CED.2	Create equations in two or more variables to represent relationships between quantities; graph equations on coordinate axes with labels and scales
	    III. Word Problems – Linear Equations
	    V. Word Problems – Systems of Linear Equations
	    XI. Word Problems – Function Graphs

A.CED.3	Represent constraints by equations or inequalities, and by systems of equations and/or inequalities, and interpret solutions as viable or nonviable options in a modeling context
	    III. Word Problems – Linear Equations
	    III. Word Problems – Inequalities
	    V. Word Problems – Systems of Linear Equations
	    V. Word Problems – Systems of Inequalities
	    XI. Word Problems – Function Graphs
	    XIV. Word Problems – Quadratic Equations

A.CED.4	Rearrange formulas to highlight a quantity of interest, using the same reasoning as in solving equations II. Literal Equations

A.REI.1	Explain each step in solving a simple equation as following from the equality of numbers asserted at the previous step, starting from the assumption that the original equation has a solution; construct a viable argument to justify a solution method
	    II. Properties
	    II. Solving Linear Equations in One Variable

A.REI.2	*Covered in Algebra 2*

A.REI.3	Solve linear equations and inequalities in one variable, including equations with coefficients represented by letters
	    II. Solving Linear Equations in One Variable
	    II. Solving Linear Inequalities in One Variable

A.REI.4	a. Use the method of completing the square to transform any quadratic equation in x into an equation of the form $(x-p)^2 = q$ that has the same solutions. Derive the quadratic formula from this form.
	    XIV. Completing the Square
	b. Solve quadratic equations by inspection, taking square roots, completing the square, the quadratic formula and factoring, as appropriate to the initial form of the equation. Recognize cases in which a quadratic equation has no real solutions.
	    XV. QUADRATIC EQUATIONS

A.REI.5	Prove that, given a system of two equations in two variables, replacing one equation by the sum of that equation and a multiple of the other produces a system with the same solutions 　　　V. Solving Systems of Equations Algebraically
A.REI.6	Solve systems of linear equations exactly and approximately (e.g., with graphs), focusing on pairs of linear equations in two variables 　　　V. Solving Systems of Equations Algebraically 　　　V. Solving Systems of Equations Graphically
A.REI.7	Solve a simple system consisting of a linear equation and a quadratic equation in two variables algebraically and graphically 　　　XVI. QUADRATIC-LINEAR SYSTEMS
A.REI.8-9	*Covered in Precalculus*
A.REI.10	Understand that the graph of an equation in two variables is the set of all its solutions plotted in the coordinate plane, often forming a curve (which could be a line) 　　　IV. Determining Whether a Point is on a Line
A.REI.11	Explain why the x-coordinates of the points where the graphs of the equations $y = f(x)$ and $y = g(x)$ intersect are the solutions of the equation $f(x) = g(x)$; find the solutions approximately, e.g., using technology to graph the functions, make tables of values, or find successive approximations 　　　V. Solving Systems of Equations Graphically 　　　XVI. Solving Quadratic-Linear Systems Graphically
A.REI.12	Graph the solutions to a linear inequality in two variables as a half-plane (excluding the boundary in the case of a strict inequality), and graph the solution set to a system of linear inequalities in two variables as the intersection of the corresponding half-planes 　　　IV. Graphing Inequalities 　　　V. Solving Systems of Inequalities Graphically
F.IF.1	Understand that a function from one set (called the domain) to another set (called the range) assigns to each element of the domain exactly one element of the range 　　　XI. Determining if Relations are Functions 　　　XI. Determining if Graphs Represent Functions 　　　XI. Function Notation, Domain and Range
F.IF.2	Use function notation, evaluate functions for inputs in their domains, and interpret statements that use function notation in terms of a context 　　　XI. Function Notation, Domain and Range
F.IF.3	Recognize that sequences are functions, sometimes defined recursively, whose domain is a subset of the integers 　　　XII. Sequences
F.IF.4	For a function that models a relationship between two quantities, interpret key features of graphs and tables in terms of the quantities, and sketch graphs showing key features given a verbal description of the relationship 　　　XI. Function Graphs 　　　XI. Word Problems – Function Graphs

F.IF.5	Relate the domain of a function to its graph and, where applicable, to the quantitative relationship it describes XI. Function Notation, Domain and Range XI. Function Graphs XI. Word Problems – Function Graphs
F.IF.6	Calculate and interpret the average rate of change of a function (presented symbolically or as a table) over a specified interval. Estimate the rate of change from a graph. XI. Average Rate of Change
F.IF.7	a. Graph linear and quadratic functions and show intercepts, maxima, and minima IV. Graphing a Linear Equation XV. Graphing Parabolas b. Graph square root, cube root, and piecewise-defined functions, including step functions and absolute value functions XVII. OTHER FUNCTIONS AND TRANSFORMATIONS
F.IF.8	a. Use the process of factoring and completing the square in a quadratic function to show zeros, extreme values, and symmetry of the graph, and interpret these in terms of a context XIII. FACTORING XIV. Completing the Square XV. PARABOLAS b. Use the properties of exponents to interpret expressions for exponential functions XII. Exponential Growth and Decay XII. Exponential Functions
F.IF.9	Compare properties of two functions each represented in a different way (algebraically, graphically, numerically in tables, or by verbal descriptions) XII. Comparing Linear and Exponential Functions XV. Graphing Parabolas
F.BF.1	a. Write a function that describes a relationship between two quantities: determine an explicit expression, a recursive process, or steps for calculation from a context XI. Function Notation, Domain and Range XI. Word Problems – Function Graphs b. Combine standard function types using arithmetic operations XI. Operations on Functions
F.BF.2	Write arithmetic and geometric sequences both recursively and with an explicit formula, use them to model situations, and translate between the two forms XII. Sequences
F.BF.3	Identify the effect on the graph of replacing linear or quadratic function $f(x)$ by $f(x)+k$, $k \cdot f(x)$, $f(kx)$, and $f(x+k)$ for specific values of k (both positive and negative); find the value of k given the graphs. Experiment with cases and illustrate an explanation of the effects on the graph using technology. XVII. Transformations of Functions
F.BF.4-5	*Covered in Algebra 2 and Precalculus*

F.LE.1	a. Prove that linear functions grow by equal differences over equal intervals, and that exponential functions grow by equal factors over equal intervals.
	    XII. Comparing Linear and Exponential Functions
	b. Recognize situations in which one quantity changes at a constant rate per unit interval relative to another
	    XI. Rate of Change for Linear Functions
	c. Recognize situations in which a quantity grows or decays by a constant percent rate per unit interval relative to another
	    XII. Exponential Growth and Decay

F.LE.2	Construct linear and exponential functions, including arithmetic and geometric sequences, given a graph, a description of a relationship, or two input-output pairs (include reading these from a table)
	    XI. Function Notation, Domain and Range
	    XI. Word Problems – Function Graphs
	    XII. Exponential Functions
	    XII. Sequences
	    XII. Comparing Linear and Exponential Functions

F.LE.3-4	*Covered in Algebra 2*

F.LE.5	Interpret the parameters in a linear or exponential function in terms of a context
	    XII. Sequences
	    XII. Comparing Linear and Exponential Functions

F.TF	*Covered in Algebra 2 and Precalculus*
G.all	*Covered in Geometry*

S.ID.1	Represent data with plots on the real number line (dot plots, histograms, and box plots)
	    IX. Dot Plots and Distributions
	    IX. Frequency Tables and Histograms
	    IX. Box Plots

S.ID.2	Use statistics appropriate to the shape of the data distribution to compare center (median, mean) and spread (interquartile range, standard deviation) of two or more different data sets
	    IX. Central Tendency
	    IX. Standard Deviation
	    IX. Quartiles

S.ID.3	Interpret differences in shape, center, and spread in the context of the data sets, accounting for possible effects of extreme data points (outliers)
	    IX. Dot Plots and Distributions
	    IX. Central Tendency
	    IX. Standard Deviation
	    IX. Box Plots

S.ID.4	*Covered in Algebra 2*
S.ID.5	Summarize categorical data for two categories in two-way frequency tables. Interpret relative frequencies in the context of the data (including joint, marginal, and conditional relative frequencies). Recognize possible associations and trends in the data. X. Two-Way Frequency Tables

S.ID.6	Represent data on two quantitative variables on a scatter plot, and describe how the variables are related
	    X. Scatter Plots
	    X. Identifying Correlation in Scatter Plots
	a. Fit a function to the data; use functions fitted to data to solve problems in the context of the data
	    X. Lines of Fit
	b. Informally assess the fit of a function by plotting and analyzing residuals
	    XI. Residuals and Correlation Coefficients
	    XI. Residual Plots
	c. Fit a linear function for a scatter plot that suggests a linear association
	    X. Lines of Fit

S.ID.7	Interpret the slope (rate of change) and the intercept (constant term) of a linear model in the context of the data
	    III. Translating "Each"
	    XI. Rate of Change for Linear Functions

S.ID.8	Compute (using technology) and interpret the correlation coefficient of a linear fit X. Residuals and Correlation Coefficients
S.ID.9	Distinguish between correlation and causation X. Correlation and Causality
S.IC	*Covered in Algebra 2*
S.CP	*Covered in Algebra 2 and Precalculus*
S.MD	*Covered in Precalculus*

Made in the USA
Charleston, SC
27 September 2014